Appleton

Applying Technology for Performance

Appleton

Applying Technology for Performance

By Russ Banham

GREENWICH PUBLISHING GROUP, INC.

© 2007 Appleton. All rights reserved.

Printed and bound in the United States of America. No part of this publication may be reproduced or transmitted in any form or by any means, electronic or mechanical, including photocopying, recording or any information storage and retrieval system now known or to be invented, without permission in writing from Appleton, P.O. Box 359, Appleton, Wisconsin 54912 except by a reviewer who wishes to quote brief passages in connection with a review written for inclusion in a magazine, newspaper or broadcast.

Produced and published by Greenwich Publishing Group, Inc.
Old Saybrook, Connecticut
www.greenwichpublishing.com

Designed by Clare Cunningham Graphic Design

Library of Congress Control Number: 2007926170

ISBN: 0-944641-74-1

First Printing: May 2007

10 9 8 7 6 5 4 3 2 1

PHOTO CREDITS:
- Page 10 (upper left) © Gianni Dagli Orti/CORBIS
- Page 11 (left) © Historical Picture Archive/CORBIS
- Pages 11 (upper right), 12 (lower right) and 22 appear courtesy of The History Museum, Appleton, Wisconsin
- Page 12 (upper left) appears courtesy of Lawrence University
- Page 36 (lower right) appears courtesy of Rosemarie De Bruin
- Pages 51 (right) and 59 appear courtesy of The NCR Archive at Dayton History
- Pages 77 (upper right), 90 (lower left) and 115 appear courtesy of Bill Van Den Brandt
- Page 113 © 2007 Journal Sentinel, Inc., reproduced with permission
- Page 130 © Lester Lefkowitz/CORBIS
- Page 139 © Adam Smith

Thanks to the many Appleton employees and retirees who contributed photographs and other archival materials for this project.

All other photographs and historical items appear courtesy of Appleton and its subsidiaries.

NCR Paper is a registered trademark licensed to Appleton. All other trademarks in this book are property of their respective owners.

Printed on Utopia Two® blue/white gloss 100 lb. text produced by Appleton Coated LLC.

Table of Contents

CHAPTER ONE:	Making Paper More Printable	8
CHAPTER TWO:	Shifting to Survive	30
CHAPTER THREE:	Riding the Carbonless Wave	54
CHAPTER FOUR:	"Get it Wet and Head it North"	78
CHAPTER FIVE:	Second Renaissance	108
	Timeline	134
	Appleton's Chief Executive Officers	138
	Acknowledgments	139
	Index	140

Making Paper More Printable

Chapter 1

The French explorer Jean Nicolet first encountered the Winnebago Indians along the Fox River, shown below in an 1858 lithograph. The tribe resided in villages of wigwams and cultivated maize, squash and tobacco, which it traded with the white man. Above, an 1828 painting of a Winnebago orator.

They came from the East in the decade before the Civil War, Yankee entrepreneurs with capital, confidence and grit to spare, drawn to the abundant woodlands and powerful waterway of Wisconsin's Fox River Valley. The swift, northerly flowing Fox plunged a breathtaking 170 feet in elevation from Lake Winnebago to Green Bay more than 35 miles away, presenting a compelling — and inexpensive — source of energy for the Easterners' paper mills.

Since time immemorial, Indian tribes like the Fox, Oneida, Menominee and Winnebago had coursed this treacherous waterway in their canoes. In the 17th century, French explorers searching for the Northwest Passage grasped the commercial value of the rapids and began building the first trading posts along its banks. The Treaty of the Cedars, negotiated with Menominee Chief Oshkosh in 1836, opened the valley to settlement by Europeans, and families from the East soon settled the region, many taking root in villages like Appleton, the largest of Wisconsin's Fox Cities in Outagamie County today.

They arrived "on foot, by boat and canoe, wagon and ox cart," Ellen Kort writes in her book, *The Fox Heritage*. "As soon as a cabin was built, it was open to those who needed a place to stay in a strange land."

Agriculture quickly blossomed, as did dairy farms sprouting local cheese factories that are still world-renowned. The first Easterner to build a saw mill was Boston merchant Amos Lawrence in 1848. Six years hence, the Richmond Brothers Paper Mill, Appleton's first, was established. The rapids were harnessed by locks, and the Fox was connected to the Wisconsin River by a series of canals. Additional links to the Mississippi and Ohio rivers made shipments possible from Green Bay to Pittsburgh, a distance of more than 750 miles as the crow flies. Virtually overnight, as author Edna Ferber wrote, the Fox became "a willing beast of burden, a benign giant whose power turned mill wheels, energized factories, created industry, (and) brought prosperity." Ferber grew up in Appleton in the early 1900s, worked as a reporter at the *Appleton Crescent* newspaper and went on to become a Pulitzer Prize-winning novelist.

The 35-mile stretch of the lower Fox River flows into the Bay of Green Bay in eastern and central Wisconsin. The river's fast current and the dense forests of Wisconsin lured paper manufacturers to the region, culminating in the building of cities like Appleton, above, circa 1900. The painting *Mouth of the River*, left, by Karl Bodmer depicts the river pre-industrialization.

LAWRENCE UNIVERSITY

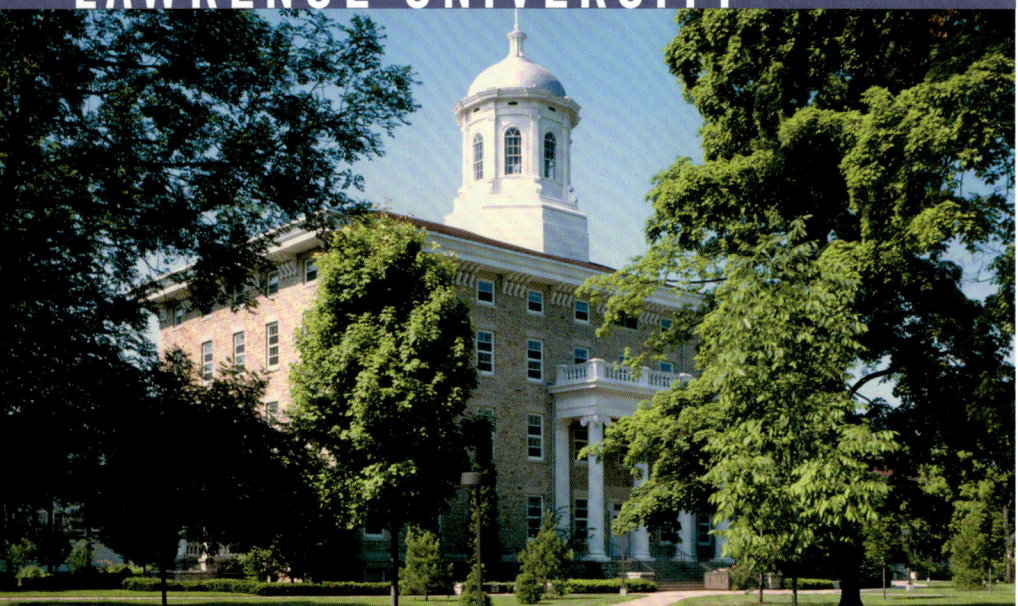

Charles Boyd was the first Appleton president to graduate from Lawrence University in the heart of Appleton, but he was not the last. John Reeve and Tom Busch also called Lawrence their alma mater. The university was chartered in 1847 and built on a wooded bluff above the Fox River. Wealthy Boston merchant Amos Lawrence and the Methodist Church of Wisconsin established the frontier school to afford "gratuitous advantage to Germans and Indians of both sexes," the university's charter states. Lawrence pledged $10,000 to endow the college on the condition that the Methodists match the gift. The Territorial Legislature granted the charter, and classes commenced November 12, 1849.

Over the years the school changed its name from Lawrence University at its inception to Lawrence College in 1913 and then back to Lawrence University in 1964. In 1929 the Institute of Paper Chemistry was founded on the campus and staffed with the world's best paper and coating scientists. Associated with the university, the institute granted both masters and doctoral degrees in paper production and coating sciences, and did specific project research by contract for individual members like Appleton Papers. In the 1970s Appleton leased space at the institute's laboratory for members of its technical department.

The three company presidents who graduated from Lawrence University are members of the Paper Industry International Hall of Fame. Busch was inducted in 1995, Boyd in 2001 and Reeve in 2004.

Lawrence University is a private undergraduate college chartered in 1847. The first classes were held on November 12, 1849. Lawrence was the sixth U.S. college to be founded as a coeducational institute. Today it resides on a 425-acre campus.

Settling the Paper Valley

Amos Lawrence had more than the almighty dollar in mind when he settled the region. With financial assistance from the Wisconsin Methodist Church, the timber baron purchased land in Appleton upon which to erect the Lawrence Institute (today's Lawrence University), a "university in the wilderness" chartered in 1847 that grew along with the town. The following year, the college's first graduating class included Samuel Boyd, a future attorney and judge whose son, Charles, would become a principal catalyst of Appleton's industrial growth a generation later.

The small town officially became a city upon its incorporation in 1857. Fed by the forests of Wisconsin and Michigan, paper mills grew throughout the land. Kimberly-Clark, Neenah Paper Company and George A. Whiting Paper Company planted flags along the shores of the Fox, and as business boomed, the Fox River Valley was nicknamed the "Paper Valley."

Seven thousand people lived in Appleton by 1880, many of them hard-working Irish, German, Dutch, English, Polish, Jewish and Italian immigrants who arrived in hordes and soon outnumbered the Eastern Yankees. The newcomers tilled the land, felled its trees and manned the mills. They opened dry goods shops and hardware stores, and the spires of their sundry places of worship pierced the sky but did not divide them. In building Appleton's first synagogue, Reformed Temple Zion in 1879, people of many faiths pitched in with money and their labors.

Rabbi Mayer Samuel Weiss, who'd arrived the previous year from Budapest, led the temple. His son, Erich Weiss, four years old when the temple was built, adopted the stage name Harry Houdini as an adult and thrilled the world with death-defying escapes.

Equally astonishing was the city of Appleton itself. In 1877 the first

telephone in service in Wisconsin connected the home of banker Alfred Galpin Jr. in Appleton with his bank on College Avenue. A telephone line was strung to a neighboring drug store and then to the office of Dr. John T. Reeve, creating one of the first telephone exchanges in the United States. The city also claimed the world's first hydroelectric central station, built in 1882 to power two paper mills and the Henry James Rogers home, the first house in the world lighted by electricity from a central hydroelectric station using the Edison system. Appleton even boasted the country's first commercially successful electric streetcar, which began rolling on January 14, 1886.

Despite these modern amenities, the predominant means of transport around the fertile river valley was still by horse and buggy. Fewer than 100 automobiles plied rutted U.S. roads and byways in the 1890s, and pedestrians were more likely to encounter oxen and carts than motorcars in the muddy streets.

The paper industry was the area's main source of investment and income. Successful paper executives cut dashing figures in their sack suits, striped shirts, removable collars, scarf ties and elastic braces. Homburg hats, walking sticks and fleece-lined underwear, a necessity in Wisconsin's frigid winters, finished the picture. Women adorned themselves in the Victorian ideal — elaborately feathered hats, voluminous skirts, puff-sleeved blouses and uncomfortably tight corsets, the latter to evince the favored "hourglass" figure.

At the end of the century couples promenaded during Appleton summers under giant elms and arching oaks, which shaded the streets into what Ferber, in her autobiography *A Peculiar Treasure*, called "cool, green naves." "At the least provocation Japanese lanterns burst into bloom on a hundred lawns, and lemonade punch bowls were encircled by organdie-clad girls and boys in white duck pants and blue serge coats," she wrote. "There never was such a town for sociability."

Paper Bag Cholly

It was in this milieu that Charles S. Boyd, son of Lawrence graduate and Outagamie County Judge Samuel Boyd, acquired the character that guided his founding and stewardship of The Appleton Coated Paper Company. Born in Appleton on November 27, 1871, Charlie, as his family called him, was raised north of Stockbridge on the family's Wisconsin farm. As a young adult, he tended his father's cows, which grazed in the family pasture on the south side of the Fox River, near the present site of Riverview Country Club. On warm summer days he swam across the cool current to the other side and back, testing his stamina and his family's constitution.

Samuel was an educated man who instilled a lifelong love of scholarship in his children, which included

Charles Samuel Boyd, shown in his 1893 yearbook photo, followed in his father's footsteps and attended Lawrence University. In the yearbook, its cover shown below left, he was remembered by classmates for his "curly locks, fair complexion and graceful deportment [making] him an object of admiration among the girls and the envy of the boys." Charlie also was extolled for his "propensity for money making … selling rags, scrap iron, hoopskirts, etc." Opposite, illusionist Harry Houdini lived in Appleton. Today a number of city landmarks — from a town square to a bar — carry his name.

Paper Bag Cholly

Charles Boyd came of age in Appleton at a time, 1893, when papermaking was listed as the city's principal industry, with more than $4 million of capital invested in the business. Wealthy paper barons were easily spotted in the city, their walking sticks denoting men of means. Small wonder why Charlie chose paper coating as his principal enterprise.

On November 27, 1871, Charlie was born to Samuel and Cornelia S. (Bowen) Boyd in Appleton. Charlie's father was an attorney and later an Outagamie County, Wisconsin, judge. English by birth, Samuel was the son of Major Thomas Boyd, an officer in the British army who with his young family had journeyed to America in the early 1840s. The Boyds found their way to the eastern shores of Lake Winnebago, where Thomas purchased land to operate a farm.

The same rugged individuality and entrepreneurial spirit coursed through Charlie. Attending his father's alma mater, Lawrence University, in the 1890s, Charlie hawked rags, scrap iron and hoop skirts, in addition to the paper bags and wrapping paper that earned him the nickname "Paper Bag Cholly," bestowed by local merchants. Following his graduation and after a few years' apprenticeship at area paper companies, he set out on his own, establishing the Charles S. Boyd Paper Company in 1905, followed by The Appleton Coated Paper Company in 1907.

Intensely involved in building Appleton Coated, "Charlie nearly forgot to marry," says Bill

Charlie Boyd founded the Appleton Coated Paper Company and was active in its leadership until his death in 1952. An amateur horticulturist, he had the factory's landscape planted with lush gardens, opposite. His beloved Packard is visible in the garage near the main office.

Siekman, his son-in-law. In 1919 Charlie was 48 years old when he exchanged wedding vows with Lois Hill Spencer, a divorcée from South Carolina whose former husband had been a professor at Lawrence. Their son, Lyle Spencer, was just a boy when his father moved to Oregon to teach. Charlie and Lois had one child, Martha, whom Bill Siekman married in 1946.

Charlie reared Lyle as his own and was extremely proud of his stepson's business accomplishments. Lyle was the founder of Science Research Associates, a Chicago-based publisher of educational materials. In 1964 he sold the company to International Business Machines in exchange for IBM stock worth $143 million. "Lyle's shares were larger than any of the sons of Thomas Watson Sr. (the first president of IBM)," says Siekman, who became chairman of Appleton Coated in the late 1960s.

Charlie was a refined man known for his taste in architecture, art and fashion. He also was an amateur horticulturist. In 1931 he built a historic estate called Villa Woodbine in Coconut Grove, Florida. The Mediterranean Revival house is revered for its fine detail and craftsmanship. "Charlie had style," says Frank Sanders, the company's human resources manager in the 1980s. "He'd wear spats and vests to work in the '50s, which by then looked out of date, but he didn't care."

The Appleton plant also didn't jibe with tradition, boasting pristine pockets of greenery planted by Charlie's gardeners. "He loved the flowers and landscaped area next to his office and alongside the mill bordering on Meade Street," recalled John Reeve, a former Appleton Coated president. "It was probably the most beautiful garden in Appleton."

The Appleton Coated Paper Company opened for business in 1907 in Appleton, close to area paper mills that would provide the base paper stock for its coating operation. Photographs of the original office building show its well-appointed gardens in the early 1900s. Left, Appleton Coated's employees in 1911.

Charlie's younger siblings Robert, Bertha and Florence. Although alcoholism eventually got the best of him and made Charlie temperate for life, the judge instilled a tenacious work ethic in his older boy. While attending Lawrence University from 1888 to 1893, Charlie's diligence hawking paper bags and wrapping paper to local merchants earned him the nickname "Paper Bag Cholly." The school's yearbook opined upon his graduation, "C.S. is a good student and business man."

On his own in the world, Charlie found a job as a paper salesman for Moser Paper Company in Chicago. He soon returned to Wisconsin to work at the George A. Whiting Paper Company in Menasha, founded in 1882 and still in operation today. For several years he toiled at the company's mill, saving every penny he earned, planning someday to invest it in a paper company of his own. To bypass the nickel fare on the trolley, he is said to have walked from the family home in Appleton to his job in Menasha, a distance of seven miles each way.

After 12 years of apprenticeship, Charlie had accumulated enough money to establish the Charles S. Boyd Paper Company in Kaukauna, Wisconsin. The enterprise was not a paper mill; rather, it was predicated on improving the printability of paper. In a rented building, assisted by eight to ten workers, the eponymous company sized base paper stock with starch or animal glue. The sizing filled surface pores, giving it a slick finish that accepted printing better. Charlie also invested in a supercalender, a machine that presses paper between polished metal and cotton rolls to increase its smoothness and gloss.

The competition was bruising. Integrated pulp and paper mills also manufactured sized and supercalendered paper, and to thwart the upstart company they denied it low-priced, pooled shipping privileges. After one year of operation, the Boyd Company struggled to survive. Charlie then had an inspiration. More sophisticated printing techniques involving the use of halftones, gradations of tone in a single color that give the effect of shading, were coming into wide usage. To produce the halftones in the printing process required paper of much higher quality than the Boyd Company produced. The base stock had to be coated with a mixture of clay and a milk protein called casein, and then dried.

Paper machines at the time could not make and coat paper in a single operation, creating a market niche for independent paper converters that could coat the paper. While nearly all converters were based in the East, a Midwestern coating operation could tap the region's paper mills for base stock. Moreover, coated papers would be accepted by the mills as non-competitive, unlike the Boyd Paper Company's sized and supercalendered papers, promising sufficient immediate demand to support operations. Charlie quickly grasped the tremendous profit potential of a Midwestern company producing coated paper in great volume.

His mother, Cornelia S. (Bowen) Boyd, agreed to invest $6,000, all the money the family had, in the enterprise,

Appleton's articles of incorporation from 1907 carry the signatures of incorporators Charles Boyd; Curtis Bynum, the company's secretary; and John Lowe, who became secretary in 1909 upon the sale of Bynum's shares.

provided that Charlie promise to financially care for his brother and sisters. He agreed. Articles of Organization for the company were placed on file with the State of Wisconsin on May 7, 1907, and recorded in the Register of Deeds office in Appleton two days hence. On May 13 Wisconsin's Secretary of State affixed his signature and seal granting incorporation. The Appleton Coated Paper Company was officially in business.

Charlie's fellow incorporators included his brother Robert, and John Lowe and Curtis Bynum, the latter men investing the remaining $9,100 in capital needed to launch the business. Charlie became president and treasurer of Appleton Coated, Robert vice president and Bynum secretary. Lowe, formerly the superintendent of the Boyd Paper Company, replaced Bynum as secretary in 1909 upon the sale of Bynum's shares in the company to his associates.

The men chose Appleton as the location for the coating plant, given its close proximity to paper mills like Bergstrom Paper Company in Neenah and Kimberly-Clark in Kimberly, which could economically deliver the base stock by horse and wagon to the Appleton plant for conversion into coated papers. The plant itself was a rented two-story, wooden-frame building north of the railroad tracks on Meade Street that had formerly housed a pea canning company. Neighbors included The Hayton Pump and Blower Company and a defunct company that had manufactured double-powered windmills. The founders renovated the 7,100 square-foot building and filled it with mostly second-hand equipment — two small coating machines, a supercalender, a cutter and a trimmer. John Lowe, the plant's superintendent through 1923, oversaw the hiring of 30 employees.

The First Successful Products

Production averaged 10,000 pounds of coated paper per day, barely enough to pay the bills. After a year of high hopes, the company's capital account showed a paltry $16,890.17. Charlie solicited $25,000 in credit from the First National Bank of Appleton and other financial institutions in the city. Banks provided the money based on his chief (and really only) asset — character. Friends also chipped in, endorsing bank notes on his behalf, and Charlie borrowed additional funds from the Boyd

Calendering is a process in which paper is pressed between a series of highly polished metal rollers, improving the smoothness and providing a better printing surface. As shown in this 1920s photo of the company's calender department, the calendered rolls were then sent to the cutter room for processing into different lengths and widths.

The Appleton Plant

Over the past 100 years the Appleton plant has sprouted on land once owned by a company that manufactured horse-drawn carriages, wagons and sleighs on the grounds. Frank Calmes built the first factory on Wisconsin Avenue in 1901 to make the vehicles, which were built on the bottom floor of a two-story wooden structure and then taken up a ramp on the side of the building to the second floor for painting. When automobiles came on the scene a few years later, Calmes Carriage Works went under, and the building was turned into a pea canning factory.

Charlie Boyd rented the two-story, wooden-frame building on Wisconsin and Meade streets to house The Appleton Coated Paper Company in 1907. He'd chosen Appleton as the site for a plant not because it was his hometown but because of the paper mills dotting the shores of the Fox River that could feed base stock to his new enterprise.

Equipment at the first plant was primitive and mostly second-hand: a couple of small coaters, a

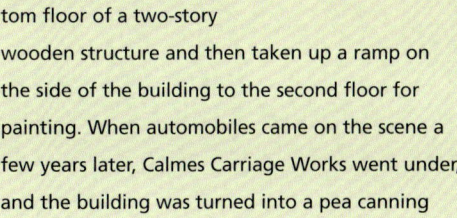

The cutter room at the plant, shown on September 14, 1928, is where men like Gus Moll, Sy DeGroot, John Brown and William Schinke, left to right, trimmed calendered paper to customer specifications. Above, the plant in 1917.

supercalender, a cutter and a trimmer. Coating was made from casein, a milk product, mixed with kaolin clay mined in England and called "English china clay" or "English coating clay," according to History of Outagamie County, published in 1912. The book estimated that the plant required 700 pounds of casein each day, putting rather severe demands on the region's cows, which also fed Wisconsin's numerous cheese factories. "It is calculated that 100 pounds of milk will make three pounds of casein, [requiring] about 1,000 cows" a day to meet the company's needs, the book noted.

Shadowing the company's relentless growth, the Appleton plant expanded rapidly and inexorably. The company regularly invested profits to enlarge existing buildings, purchase real estate, erect new buildings and introduce more modern equipment. Every passing year revised the plant's boundaries, which pushed out north, south, east and west. To accommodate the need to expand the plant in the 1960s, the company made an unusual request, successfully petitioning the city of Appleton to allow it to move the north part of Meade Street 150 feet to the west. More buildings soon filled what once was the street.

Today the Appleton campus comprises more than a dozen buildings dotting both sides of the railroad tracks. The Appleton plant is listed at 825 East Wisconsin Avenue, but then there is the Rankin Street plant at 1020 North Rankin. The Meade Street laboratory is located at 1100 North Meade, while the Hancock Street lab is at 714 East Hancock and the Lawe Street lab is at 1100 North Lawe. A wellness center named for former president and CEO John Reeve is at 1111 North Rankin, in the same building as the No. 17 thermal coater. "We actually slid the wellness center in under the disguise of the new coater," says Dale Schumaker, Appleton CEO in 1993. "It never showed up in the official plans. We had a little bit of money left over and wanted to do something for employees. When we dedicated the center in 1996, John was thrilled."

Altogether, the Appleton plant encompasses more than one million square feet of space; approximately 800 employees work at the plant under the supervision of plant manager Paul McCann. Although Calmes Carriage Works is a distant memory, relics of his vehicles are said to be evident throughout the state.

Students at Lawrence University, below, celebrate the university's victory in an oratorical contest on May 3, 1900, surely surpassing the capacity of Appleton's College Avenue streetcars. Appleton plant employees in the early days, opposite, check the finish of freshly calendered paper.

Paper Company, which moved to Appleton in 1910. The two firms shared plant and personnel but kept separate accounts until the older company went out of existence 30 years later.

The capital infusion helped Charlie acquire the rented buildings and adjacent land, pay salaries and other operating costs, purchase new equipment and build additional facilities. The latter included a wooden structure in 1909 that contained a small room for mixing colors in used oil barrels and tanks, and a modern, four-story warehouse erected the following year. Increased demand for coated papers compelled the need for additional employees, and 60 workers manned the plant by 1910, at an annual payroll of $27,000. One of the new hires was an ambitious 16-year-old named Fred Heinritz, who ran errands.

The company's first successful product was White Porcelain Enamel Shelf and Lining Paper, introduced in 1911 and sold under the private brand names of paper wholesalers and merchants. Charlie solicited the business personally, traveling by train to meet with middlemen called jobbers, who sold his products to the paper merchants. The orders were coated according to each customer's specifications.

To make coated paper, kaolin clay imported from England — it was commonly known as English china clay — and water were placed in a tank and stirred with a hand paddle until adequately dispersed. Casein obtained from Wisconsin dairies in large, hardened curds, was ground up and added to this mixture as an adhesive to help bind the coating to paper. After stirring and straining, the resulting brew was poured into a coating machine using a hand dipper. The machine applied the coating to the base paper using circular brushes, and the coated stock was then hung above steam coils to dry. This festooning process, as it was called, limited production to a mere 150 to 200 feet of paper per minute. A marginal increase in speed was achieved in 1915 with the addition of an industrial fan to circulate air.

After drying, the coated paper was fed through a super-calender to impart smoothness and gloss and then dispatched to the rewinder department, which slit and rewound larger rolls of coated paper into smaller finished rolls. Depending on the order, rolls might also be sent to the sheeting department for processing into different widths and lengths, and/or to the trimming department, where trimmers cut sheets to customer specifications. The final product was then packaged in the box room for shipment to customers. A waterproof lining in the box box protected the contents. "To the people of Appleton, it was an innovation to have an industry of this nature in its midst," *The Post-Crescent* of Appleton marveled in a 1922 profile of the company.

The coated papers were distributed principally in the Midwest, and their relatively good quality afforded repeat business from merchants like The John Leslie Paper Company of Minneapolis, Kansas City Paper House, Dwight Brothers Paper Companies of Milwaukee and Chicago, and Topeka Paper Company, among others. Charlie's goal was national distribution, and over time the company attracted paper wholesalers from San Francisco to New York.

Frugal Charlie's Company Prospers

Although he worked non-stop, living a bachelor's existence in a modest home on East Crescent Street, Charles Boyd become a well-known figure around town, a tall, imposing gentleman highly regarded for his taste in clothing and refined manners. "Charlie had learned deportment at Lawrence University," explains Bill Siekman, a former Appleton chairman who married Charlie's daughter Martha in 1946.

A popular story has it that in the early days of the plant, back when the company generated most of its electrical power from steam, it was customary to blow a steam whistle at the beginning and end of each workday. When Charlie learned that the whistle consumed $12 worth of steam each time it blew, he quickly ended the practice.

At the plant the boss was better known for his frugality, constantly urging employees to devise ways to conserve operating and production expenditures. A popular story has it that in the early days of the plant, back when the company generated most of its electrical power from steam, it was customary to blow a steam whistle at the beginning and end of each workday. When Charlie learned that the whistle consumed 12 dollars' worth of steam each time it blew, he quickly ended the practice.

Streetlights came to Appleton in 1912, as did a major winter storm that jammed the Fox River with ice, flooding the power plant and knocking out the steam turbine in minus-22-degree weather. A more formidable challenge emerged the following year, when the company fell behind in payments to Kimberly-Clark, a major supplier of base paper. "Mr. Boyd told me he nearly went under, and was sweating blood," said John Reeve, who joined the company in 1934 as a sales assistant and whose grandfather, Dr. John T. Reeve, was among the first people in the city to have a telephone.

"Frank J. Sensenbrenner, then president of Kimberly-Clark,

The Appleton Coated Paper Co.
APPLETON · WISCONSIN

Research and development was a critical part of The Appleton Coated Paper Company in the early years, as it remains today. The plant's laboratory was adjacent to the mill office, shown left on a spring day in 1951. The 1917 photo of the plant, opposite, features a relic of the past in the form of a horse and buggy. The sign, "Look Out for The Cars," at the railroad crossing indicates the proliferation of the automobile. Always a fast if not reckless driver, Charlie Boyd barely missed the train screaming along the tracks on his way to work in the morning.

came personally to discuss the serious matter with Mr. Boyd," recalled Reeve, whose father, Howard, was one of Charlie's closest friends. "At the conclusion of the meeting, credit was extended and catastrophe was avoided. But it was a very close call." The credit kept the company in the black and also helped fund the construction in 1914 of the Main Building, as it was called, at a cost of $44,000. The new structure housed the coating, calendering, sorting, cutting and trimming departments.

Appleton Coated prospered in the United States as war was brewing in Europe. Annual tonnage increased sharply from 1,375 tons in 1910 to 2,630 tons five years later, and in 1916 the company recorded its largest profit to date. By the end of the decade, it led the industry with the widest range of coated papers in the country.

The Roaring Twenties, the decade-long economic boom ignited by increased industrialization, began with a bang for Appleton Coated. Sales crested the $2 million mark in 1920, a 205 percent increase from the previous year. The company shipped 6,800 tons of coated papers, 43 percent higher than 1919's tonnage and substantially higher than the 1,375 tons recorded in 1910. Much of the tonnage increase came from several large orders, allowing the company to produce paper in longer, continuous runs on its two coaters. These long runs increased production efficiency while strong demand pushed selling prices upwards, and profits reached a record $230,000 by the end of 1920. Two years later *The Post-Crescent* reported that the company was coating more than 50,000 pounds of paper a day.

As profits steadily increased, the company plowed them into operations. It erected a $54,000 trimming and packaging building in 1920 and enlarged it two years later for an additional $65,000. The same year, Appleton added a second warehouse, at a $33,000 cost, to the factory complex. Employees earned a good wage and in 1920 received the first group life insurance plan sponsored by an industrial firm in Wisconsin. This close-knit group of 150 to 175 workers formed a company-sponsored baseball team, with white and blue collars interspersed in the batting order. "The team did very well in the '20s," Reeve recalls.

Many returning war veterans had joined the ranks, including the company's first chemist, secured to develop new coatings that better matched customer needs and to troubleshoot the ones that failed to work properly. Many modern operating processes made their debut, including a production accounting system that measured expenses in each phase of the coating process. The system reduced waste, enhanced efficiency and permitted sales people to quote more competitive prices to customers.

Orders continued to rise through the decade, compelling greater supplies of raw materials like clay and casein. Each month, Appleton Coated bought more than 60,000 pounds of casein, "for which two million pounds of skim milk was necessary," *The Post-Crescent* observed. Most of the casein was imported from Argentina because U.S. dairy producers could supply "only half the trade," the newspaper noted.

The Product Line Expands

Confident that his company was on the road to success, Charlie finally succumbed to marriage in 1919, at the age of 48 years. His bride, Lois Hill Spencer, had been previously married, and Charlie took her eight-year-old son Lyle under his wing, rearing him as his own. The newlyweds moved into a house on East Lawrence Street next to Charlie's mother's home and bought a small dairy farm and cottage two miles away, overlooking the Fox River. In 1921 the couple enjoyed the birth of their only child, Martha.

Charlie still ran the company on a daily basis but began to delegate more work to his lieutenants like Fred Heinritz, who'd risen through the ranks from his days as an errand boy to become a member of the board of directors in 1920. Other key executives included Matt D. Weyenberg, who led the bookkeeping and accounting departments, and Richard Mahony, who ran the cost department. Although Charlie's brother Robert was vice president of Appleton Coated, it was a figurehead position. "Things just didn't work out for Robert at the company," says Reeve. "As a young man, he would occasionally be sent home by Mr. Boyd for a week for acting up in the office. He was far too fun-loving and carefree … as different as night from day from his brother Charles."

The younger Boyd spent much of his career as a salesman

The 1920s roared at The Appleton Coated Paper Company. In 1920 the company employed 151 people at a payroll of $200,000, a far cry from the 60 employees and $27,000 payroll listed 10 years earlier. Income was continually reinvested in new buildings and machinery, including a new trimming and packaging building in 1920 and a new finishing room, circa 1922, below. Plant workers formed a company-sponsored baseball team that fared well against other company teams. Left, the 1924 squad poses for its official portrait, the team logo "ACPCO" emblazoned on their uniforms.

for the Moser Paper Company in Chicago, though he remained an Appleton Coated board director and vice president until his death in 1949. Charlie never reneged on his promise to his mother to take care of his siblings.

In 1922 the company's foremost products were listed as "enameled book and magazine papers, coated on both sides, and lithograph labels, coated on just one side and used for high-end illustrating, magazines and catalogs." The following year, a new product, Coated Bond paper, joined their ranks, developed for printing applications using gelatin roll duplicating equipment, a new photographic process. Coated Bond was the first paper in the market for this purpose; the company produced vast quantities and sold it to jobbers.

Out of this specialty grew the company's first mill brand line — its first product under its own brand name. Called Direct Sales Bond, this coated-on-one-side bond paper was perfect for the four-page illustrated letter, a new direct-mail advertising format coming into wide usage at the time. The non-coated bond surface of the paper was used for the advertisement's letterhead and correspondence, while the coated side, which boasted improved opacity and printing surface, held the illustration. The bond side was available only in white, and the coated side was available in five colors — white, buff, green, blue and goldenrod. "[Direct Sales Bond] has excellent printing qualities and will readily take halftones as fine as 133 line screen," the company pointed out in its first national advertising.

Direct Sales Bond found a ready market, and within a short period of time was carried by 90 distributors in the United States, Canada, the Dutch East Indies, Australia and New Zealand, marking the first Appleton Coated product to be sold internationally. It quickly was followed by other mill brand lines, most for use in the production of direct-mail advertising. Many represented bold departures from the norm, different from anything else on the market. Such was the case with Woodbine Colored Enamel, which capitalized on the growing preference in the marketplace for brilliant colors, as opposed to the pastel shades used in Direct Sales Bond.

Woodbine Colored Enamel was the first in a long series of high-quality, brightly colored papers manufactured by the company, and it soon became the world leader in this market. The product's unusual name derived from the Woodbine ivy (also known as Virginia creeper) growing along the walls of the plant. Charlie, an amateur horticulturist, had planted the ivy and many other plants, shrubs and flowers to soften the factory's institutional feel.

Another groundbreaking product followed — Polychrome Dull Coated Book. The prevailing wisdom at the time was that dull coated book would not print well since it was not calendered. Several experiments at the plant proved otherwise. "Appleton Coated was less impressed than other converters by the distinction between the possible and the impossible," the company's half-century brochure, *50 Colorful Years*, boasts.

Two hundred employees worked at the company, led by Heinritz, who became the general manager of the plant in 1925. "My dad learned the coating business on the job, and furthered his education by taking classes at Bowlby Business College downtown," said Fred "Bud" Heinritz, who joined the company in 1947 and retired as a national account sales manager in 1987.

Thanks to growing demand for the mill brand lines, Appleton Coated's total volume in 1925 exceeded 8,000 tons. To accommodate the surge in orders, the company built a second warehouse and a $133,000 addition to the Main Building to house the calendaring, cutting and sorting operations. It also acquired and rebuilt the neighboring Hayton Pump and Blower Works building, at one time a machine shop, foundry and pickle factory, to prepare dyes and other coating ingredients. By the end of the decade, Appleton Coated operated eight brush coaters at the plant.

Coping with the Depression

A boon to Appleton's business came with the founding in 1929 of the Institute of Paper Chemistry, an independent, privately supported graduate school. Via a close association with Lawrence College (Lawrence called itself a college rather than a university between 1913 and 1964), the Institute offered master of science and doctoral degrees in paper chemistry. Appleton Coated became a member of the Institute and in succeeding years often turned to the school's world-renowned scientists to obtain expert opinions on a wide range of coating problems and challenges, and new coating processes.

The Roaring Twenties whimpered to a close on October 24, 1929, when panic selling of stock on Wall Street sent the Dow Jones Industrial Average plummeting. By November the Dow had fallen from 400 to 145, erasing more than $5 billion in share values in three days. When the reverberations calmed, more than $16 billion in stock capitalization had vanished. A massive economic recession gripped the country.

The Great Depression challenged Appleton Coated with the tightest markets in its quarter-century of life, yet the economy was only part of its troubles. Early in the decade, papermaking machines had been redesigned to both make and coat paper in a single operation. The innovation created stiff competition for converters like Appleton Coated, which were forced to relinquish certain grades of coated paper to the less expensive mass-production techniques used by the integrated paper mills. Over the next few years, Appleton Coated phased out one white coated-grade product after another. "Everybody thought (we) would go under," Richard Mahony recalls.

Few men, especially one in his 60s, would have the fortitude to stay the course. But Paper Bag Cholly had another ace up his sleeve. He confronted the loss of his primary business by radically shifting gears to produce specialty papers, created and coated for functional purposes other than just printing. While Appleton Coated had manufactured such papers in the past, they were a marginal source of revenue.

Now Charles S. Boyd bet his company's future on its secondary line — in an era when most men were lucky to have a job. As he often remarked about times of duress, "Your true colors show."

Direct Sales Bond, opposite, was the company's first mill line under its own brand name. The paper was coated on one side and used in new direct mail advertising applications. Fred Heinritz, left, who worked for Appleton for 40 years, was general manager of the plant in the mid-1920s.

Shifting to Survive

Chapter 2

Charlie Boyd was 63 years old in 1934, an age when most men blissfully plan their retirement. The stock market crash forbade such contemplations, spawning a massive industrial recession that undermined the foundations of Western capitalism. In just five years, U.S. industrial stocks had lost 80 percent of their value. The gross national product was down by a third and international trade by two-thirds. More than 10,000 banks had failed, causing more than $2 billion in deposits to vanish. One out of every four workers was out of a job, and many lingered on long bread lines to feed famished families. Charlie soldiered on.

Thanks in part to family connections, John Reeve, best friend of Charlie's stepson Lyle Spencer, was among the few to secure work at Appleton Coated in 1934. "I felt extremely fortunate to have a job, as I had many friends who had been out looking for work for years," Reeve said. He was finishing up a degree in economics from Lawrence College, and to fortify his chances of employment spent months poring over books at the

Institute of Paper Chemistry. He also took a course in how to operate a printing press at Appleton's Vocational School, the first U.S. facility built exclusively for vocational education and a predecessor to Fox Valley Technical College. "I wanted to learn all I could," Reeve confided.

Appleton Coated, meanwhile, lumbered along. "The Depression years were rough," says Frank A. Sanders, a long-time Appleton employee whose father, Frank E. Sanders, worked in maintenance for the company during the Depression. "Dad's work week was cut to three days, and he, the pipe fitters and the machine shop guys were lucky if they made $10 a week to live on. But at least the company stayed afloat when so many others foundered."

Manning the helm was Charlie. As president and chairman of the board, he approved all large capital expenditures. Fred Heinritz, the company's secretary and general manager, watched over day-to-day plant operations. Matt Weyenberg, head of purchasing, and Dick Mahony, the office manager who

Most Appleton workers held onto their jobs during the difficult period of the Great Depression — and, of the 100-plus employees shown in this 1936 photograph, at least three were still on the payroll 40 years later.

Polychrome Dull Coated Book, part of Appleton's specialty papers line, was promoted in 1931 via a sample book, above, developed by Norman T. A. Munder, a leading graphic artist. A page from a marketing kit in the 1950s, opposite, promoted the benefits of printing color on colored stock. This advertisement also can be seen in the center of the display at right.

kept a sharp eye on finances, filled out the top ranks. "Fortunately, Charlie had the gift of surrounding himself with capable people," says Bill Siekman, a former chairman and Boyd's son-in-law.

During the Depression the company made the difficult switch to specialty coated papers — products requiring unusual decorative elements or meeting a specific technical need. On the bright side, there was scant competition in this market from integrated pulp and paper mills. Unlike white coated grade products produced by the mills in large volume in long, continuous machine runs, the machine runs of specialty papers were comparatively short and required significant downtime to clean up the machine before undertaking the next run. The mills balked at the uneconomical costs per machine hour, opening a clear path for Appleton Coated.

Color and Coatings

The strategic imperative for Appleton Coated in the 1930s was to exploit color and coatings in a variety of formats to garner an increasing share of the specialty market. The Institute of Paper Chemistry was invaluable in this quest. When technical problems required resources beyond those available at the company, Appleton submitted them to the Institute for analysis. This unique collaboration yielded an amazing array of specialty papers in the 1930s, helping Appleton Coated penetrate the era's drum-tight markets with several unique products.

Serendipity also intervened in the creation of the Wisconsin Paper Group, a marketing organization that grew out of the efforts of Appleton Coated and Bergstrom Paper Company in Neenah, Wisconsin. Both companies realized the value of combining the region's plentiful paper resources for marketing purposes. The organization advised distributors to purchase a majority of their paper products in the state, pool these shipments and then move the tonnage in regular train carloads, reducing their overall costs. Wisconsin Paper Group is still in operation today.

Appleton Coated's first successful specialty product was Duplex Coated Paper, which featured a bold color on one side of the paper and a harmonizing tint on the other. The paper was produced in a broad range of color combinations, such as

FASHIONS FOR THE YACHTING SEASON

orange and fawn, goldenrod and primrose, turquoise and India, and gray and rose. "With the printer's ink supplying still another color, three-color direct mailing pieces could be obtained with a single printing," the company stated.

Another winning specialty product was a heavier duplex paper than Woodbine Duplex Enamel, initially called Post Card to reference its primary application but later changed to Woodbine Duplex Coated Bristol. Also popular was label paper in a wide range of colors for cans, bottles and other merchandise, carried under the brand names Appleton Label Plate and Litho Label; and coated tag and C1S paper (paper coated on one side), used to manufacture gummed paper and blotting paper. Other specialty lines were Dullchrome Coated Book, a dull-coated, two-tone paper for direct advertising, and Mellochrome Dull Coated Bristol, heavier in weight than Dullchrome Coated, with the same color on both sides.

Appleton Coated received an unexpected assist in 1933 from the National Recovery Administration (NRA), the keystone of President Franklin D. Roosevelt's "New Deal for the American People." The NRA permitted businesses to draft "codes of fair competition" regulating prices, wages, working conditions and

A 1950 Appleton marketing kit, above, offered a "head start" to companies promoting summer tourism. "Millions will be spent this year for summer vacations," accompanying text reads. "Go after your share with folders, broadsides, booklets or catalogs printed on Polychrome Dull Coated Book." The Institute of Paper Chemistry, right, did research for Appleton and other companies.

terms of credit. Businesses that complied with the codes were exempted from antitrust laws. The NRA standardized white enamel paper products, establishing four grades with a fixed price for each. Appleton Coated capitalized on the opportunity, launching a brand line for each grade — Masterfold Enamel, Woodbine Folding Enamel, Appleton Enamel and Wisconsin Enamel.

Despite the timely boost from the U.S. government, Charlie was no fan of Roosevelt or his New Deal. One day Gene Colvin, the company's traveling salesman, returned from a trip East. Charlie asked him how things were going. "I've been calling on some new dealers," Colvin replied, in reference to distributors. Charlie's eyes narrowed. "I don't think much of the New Deal, you know," he shot back. Evidently, the salesman never could persuade Charlie that he hadn't been chatting with Roosevelt insiders.

Becoming the Largest

The creative braintrust at Appleton Coated burst with new ideas during the 1930s, daringly experimenting with different coating mixtures to supply a broad range of customer needs. Many products were groundbreaking, such as Permakolor Litho Label — the first coating for one-side label papers that did not fade or bleed when high-gloss inks or varnishing were applied in the printing process.

The company also was a pioneer in the production of coated book paper for offset printing. Previously, most magazine production utilized letterpress printing. In offset printing, the ink is transferred from a photographic plate to a rubber roller and then to paper. As the roller presses against the paper moving through the press, the image is left on the paper. Appleton Coated unveiled the first stock line of paper for offset printing, called Empress Offset Enamel, and gradually converted other specialty lines like Polychrome Dull Coated Book and Woodbine Colored Enamel for offset printing applications.

The company's technical department, which oversaw research and development, also formulated a new type of coating that included fine metallic powders for making gold and silver colored papers. The process was less expensive than using non-metallic dyes and inks and produced a much higher-quality paper for catalogs, booklets, annual reports and, as a matter of pride, the front and back covers of *50 Colorful Years*, the company's first history book. The company called the new line Currency Cover.

Another new product satisfied demand for a durable and rigid paper that could accommodate spiral, plastic and other books with mechanical bindings, which were coming into prominence. Called Supertuff Cover, the paper was one of the highest quality coated covers in the marketplace. The paper absorbed varnish, lacquers and high-gloss inks without fading or bleeding, and its firmness and versatility made it a stylish favorite for use in menus, counter displays and programs.

The biggest specialty line actually was among the least sophisticated — coated one-sided enamel Litho Label paper used for making cigarette packages. Several tobacco companies, manufacturers of Lucky Strike, Chesterfield and Pall Mall brands, used the paper to make "cups," the industry's term for the interior package containing the cigarettes. Litho Label "was strong enough to help prevent crushing the cigarettes, [with] a printing surface smooth enough for the manufacturer's trade-marked package labels," *The Post-Crescent* reported at the time.

More than 30 different paper mills supplied the company with base stock to make this expansive array of products. The various lines were hawked by Colvin and his sales assistant Reeve, who traversed the country by train. "With Mr. Colvin, I attended many direct-mail advertising meetings, associates' exhibits and conferences in different cities promoting our grades to printers, advertising agencies and large companies that had advertising departments," Reeve recalled.

In 1935 Appleton Coated shipped nearly 10,000 tons of specialty paper, representing a sales value of $1.385 million — less than the $2 million recorded in 1920 but nonetheless affording a profit of more than $60,000 after taxes. Appleton Coated had become the largest manufacturer of specialty papers in the world, meeting the test of losing its primary business during the darkest economic period in American history.

Charlie Boyd, in his 60s, was still coming into work every day at 9 a.m. sharp and making frequent visits to the plant floor in the mid-1930s, putting on an old gray sweater with patches on the sleeves to make workers feel more comfortable in his presence. Here, in one of the few photographs available of him in his later years, he appears in more formal — and, for him, more typical — attire.

Richard Mahony, above, was Appleton Coated's office manager in the 1930s and '40s. He later became company president and successfully changed the company's strategic course. Casein, a milk protein that Appleton used as a binder for its specialty coated papers, was delivered to the plant in sacks, below. A job in the color department was messy work, as employees in 1954, opposite, in "Safety First" hats and splattered overalls would surely testify. Left to right, Leo Hughes, day color mixer; Lloyd Mittelstadt, tank loader; Harold Radke, helper; Carl Lust, tour boss; Woodward Wheelock, strainerman; Carlton Stark, color mixer.

Close-Knit Atmosphere

Charlie had kept as many employees on the payroll as he could and made new hires when business demanded. Some new employees found work through interesting circumstances, including a penniless man who had surreptitiously moved into a vacant frame house the company owned on Wisconsin Avenue, east of the finishing building. "When Mr. Boyd was told of this, rather than just eject the man, he gave him a job in the mill," Reeve said. "In this way the man could retain his dignity, use the house and afford to pay rent."

The boss came to the office every day at 9 a.m. sharp. "Often, he would walk to and from work from his home on East Lawrence Street," Reeve said. "While at the office he enjoyed reading the papers, smoking either a pipe or cigar, and visiting with people. He would occasionally tour the mill. He loved the flowers and landscaped area next to his office and alongside the mill bordering on Meade Street."

Siekman says that whenever Charlie took a jaunt through the plant, he'd don an old gray sweater with worn-out patches on the sleeves "to make the hourly workers feel comfortable in his presence. From someone who preferred the era's more elegant clothing, the thoughtful gesture was appreciated."

Appleton invested heavily in machinery and plant improvements after the war. Opposite, an employee operates one of the company's trimmers, which cuts huge paper rolls to size for customers. The company also invested in its people, offering a pension plan in 1947. In the early 1940s, as these employee handbooks show, factory workers were paid wages of between 40 and 61 cents an hour, depending on length of service, with women making about a dime less than men for the same work.

ARTICLE IV

Wages

Section 1. Effective as of February 15, 1942, the hourly wage rate of all employees shall be increased by four cents (4 cents) per hour, and the minimum wage scale shall be established as follows:

	Rate Per Hour	
	Male Employees	Female Employees
Probationary period	54c	40c
After two (2) months' service	55c	45c
After six (6) months' service	58c	48c
After twelve (12) months' service	61c	50c

Section 2. When an employee is promoted to a job paying a higher rate, the increases shall be made in the following manner:

Rate Difference Old Job and New Job	Rate Advance at Change	Rate Advance 3 Months	Rate Advance 6 Months
1c	1c	0	0
2c	2c	0	0
3c	3c	0	0
4c	3c	1c	0
5c	3c	2c	0
6c	3c	2c	1c
7c	3c	3c	1c
8c	3c	3c	2c
9c	3c	3c	3c
10c	3c		

Section 3. Piece work... ary 15, 1942, shall be... of this agreement; prov... shall not serve to deny... request rate adjustments... agreement.

Section 4. When an... transferred from his regul... lower rate, his wage rate... provided, however, that... regular job.

Section 5. When an employee is required under the seniority procedure to take a job paying a lower rate, he shall be paid the rate applicable to such job.

Section 6. All wage schedules mutually agreed upon by the Union and the Company shall become a part of this agreement.

ARTICLE V

Seniority

Section 1. This article shall apply to all employees, except newly hired employees with less than sixty (60) days of employment to their credit.

Section 2. The seniority rights of any employee shall be measured for all time worked from the date of his first employment; provided, however, that any employee who is rehired after voluntarily quitting or having been properly discharged shall have his seniority measured from the date of rehiring.

Section 3. The Company shall furnish a seniority list by departments, showing the hiring date and seniority position of each employee. Such list shall be kept on file by the Union and shall be kept up to date at all times.

ON THE JOB AT APPLETON COATED

June 1943

YOU and YOUR JOB at APPLETON COATED

JUNE, 1942

Herman Berge worked at the company for 50 years. His father had been a carpenter with Appleton Coated, and Herman started work in 1917 as an errand boy. He went on to head the Chicago sales office and later became a member of the board of directors. In a classic Appleton photo from the 1940s, he is shown, opposite, with a pack of "Luckys," for which Appleton made the liner cups.

Executive managers and other salaried employees worked in the main office, located in an old, two-story wooden frame house. The new plant was attached to this house, which was "small and overcrowded," Reeve griped. "In the early days of telephone switchboards, everyone in the office could overhear what was said. One day the operator shouted to me, `John, Jean says bring home a loaf of bread and a bottle of milk!'"

Other employees recall hearing the frequent bellow, "Ruby!" coming from Charlie's corner office. Ruby was Charlie's secretary.

The plant was still suffused with the same close-knit atmosphere of the early days. "There was a good spirit and feeling among employees, possibly because everyone felt so fortunate to have a job," Reeve said. "When an employee had a birthday, it was his or her duty to provide a treat for everyone in the office on that day. Every year there was a Christmas dinner party with songs, stunts and white elephant gifts exchanged. Sometimes there was an office picnic at High Cliff Park."

Things were happening outside the Appleton plant as well. In 1933 the company opened a sales office in Chicago, led by Herman Berge, whose father had been a carpenter with Appleton Coated in its infancy. The company shared the office with Bergstrom Paper Company, a major supplier of base stock to the plant. Like Heinritz, the younger Berge got his start at the company as a 15-year-old errand boy, rising through the ranks to eventually become a board director.

Workers formed their first union in 1940, following passage five years earlier of the National Labor Relations Act, which protected the rights of private-sector workers to organize labor unions. Reeve was chosen to manage company relationships with the union, overall plant safety and other personnel matters, such as the hiring and training of foremen. "Up until then we had a poor safety record, with far too many lost-time accidents," he recalls. "With special effort, there was marked improvement in this area."

Charlie took the union's establishment in stride. He had always been a benevolent employer, offering a range of employee benefits well before other companies, but he also was pragmatic. "On the one hand, he thought it simpler that he no longer had to deal with employees on a one-by-one basis, since that now would be handled collectively by the union and John Reeve," Siekman explains. "On the other hand, he was dismayed that he no longer would be able to reward those individuals who worked harder and better than others."

By 1940 the worst of the Depression was behind the country. Appleton Coated survived not only intact, it was in better shape financially than most enterprises. "We were the only paper company in the U.S. not to have lost money during the Depression," said Reeve. "The company was never in the red. However, in one year the profit was a total of only $5,500."

Wartime Changes

The new decade began promisingly. Production in 1940 exceeded 11,400 tons, net sales topped $1.8 million, invested capital hovered around $1.25 million and inventory averaged $350,000, robust figures for the times. It took nearly 600 carloads to ship all the tonnage — at a $100,000 price tag.

Three hundred employees now worked at Appleton Coated, eight of them in the technical department. Led by Abe Lewenstein, a dye chemist by profession, they experimented with a wide range of coatings and dyes "in a splendidly equipped laboratory," as *The Post-Crescent* described it. The lab was responsible for the potpourri of products pouring forth. A glance at the product menu in the early '40s reveals an exquisite assortment of papers used to make postcards, shelf liner, cigarette cups, magazine covers, merchandise labels, campaign placards and even the paper imitating cork in cigarette filters. "The firm makes coverings for everything from shoe boxes to fancy candy and other high-grade boxes," the newspaper marveled.

After the United States officially entered World War II, expansion activities at the plant ground to a halt, and the company temporarily stopped making several coated and colored grades due to wartime shortages. "Casein was unavailable," Reeve explained. "A substitute had to be found." The biggest problem was that casein imports had dried up. Local dairies were restricted because most milk was needed for the war effort. Appleton Coated finally had to make its own binder from de-oiled soybean flour purchased from the Glidden Company. The protein was leached out in huge, 10,000-gallon

At the Charter Dinner for Appleton's first Quarter Century Club members, held at the Conway Hotel in downtown Appleton in 1946, attendees sang "My Wild Irish Rose," "Down by the Old Mill Stream" and other songs, whose lyrics were printed on the back of the program, right. Above, left to right in a photo from the 1947 dinner, Charlie Boyd, Fred Heinritz, Max Eggert, George Regenfuss, Eva Mielke, Emil Kahler and Vilas Dorschner gathered for a portrait.

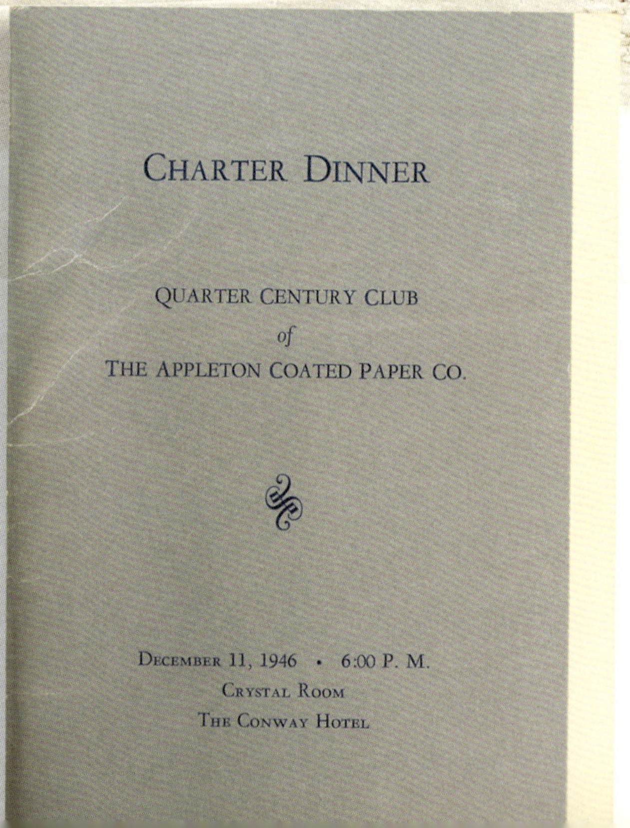

Charter Dinner

Quarter Century Club
of
The Appleton Coated Paper Co.

December 11, 1946 • 6:00 p.m.
Crystal Room
The Conway Hotel

wooden vats. Once treated, dried and ground, it became Apro, a temporary substitute for casein.

Like many other manufacturing concerns, Appleton Coated provided materiel for the war effort. The plant machine shop operated 24 hours a day and seven days a week, making tank treads, parts for submarines and the tops of eight-inch aerial bombs. The U.S. Army purchased more than a million pounds of Supertuff Cover for use in silk-screened outdoor signs and placards, and more than a million pounds of Woodbine Colored Enamel in goldenrod and orange for masking purposes in the production of offset printing plates for aerial photography and map making.

Most famously, Appleton Coated was enlisted to change the color of its coated paper for Lucky Strike cigarettes from green to white (the prominent red bulls-eye was retained in the printing process). American Tobacco Company, which manufactured the brand, claimed the switch was necessary because of limited supplies of copper used to make the green color. Throughout the war, American Tobacco advertised "Lucky Strike Green Has Gone to War," a pitch that's considered one of history's greatest marketing promotions.

Charlie finally passed the baton to Fred Heinritz in 1945, although he remained chairman of the board. The new president had been running the plant for years and had been on the board since the '20s. "My dad was a very hard-working guy who would go down to the plant on Sunday mornings to open the mail to see the orders coming in," says Bud Heinritz. "Although Mr. Boyd and my dad didn't socialize, they shared great mutual respect."

Huge Surge in Orders

Fred Heinritz took command at a fortuitous time. In 1945 Appleton Coated recorded its highest revenues to date — $2.89 million on more than 15,000 tons of paper shipped, despite a workforce that had fallen to 287 employees because of enlistment and the draft. Wages for men who remained at the plant increased from 67 cents an hour to 73 cents, and for women from 55 cents an hour to 60 cents. The women primarily worked in the sorting department, sampling papers for defects. "We had a team of about 70 ladies sorting, fanning

QUARTER CENTURY CLUB

Created in December 1946, the Quarter Century Club recognizes Appleton employees with 25 or more years of continuous service. Two employees — Matt D. Weyenberg and Herman B. Berge — made the club twice, in a sense, serving for 50 or more years. This photo shows the members of the first Quarter Century Club in their official portrait. Back row, left to right: Edward Muenster, Frank Sanders, Lee Chady, Lester Helser, George Sylvester, Frank Nowak, William Farquhar Sr., Edward Peotter and Gus Moll. Middle Row: Fred Heinritz, Lloyd Smith, Matt Weyenberg, Richard Wheeler, John Otto, Elmer Koerner, Earl Rogers, Herman Berge and Henry Ashman. Front row: Earl Helser, Ethel Denstedt, John Moll, Florian Zeffery, Anton Gauerke, Louis Thies, Cora Koletzke and Ted Lang. The inset photos are of Fred Kubitz and Charlie Boyd.

Each year the club hosts a member induction ceremony, which varies among company locations. At Appleton's 50th anniversary banquet in 1957, nearly 25 percent of club members were current employees. More than 500 current and retired employees and their guests attended the 61st Quarter Century Club dinner at the Appleton plant in 2006. It was hosted by plant manager Paul McCann, an inductee himself that year.

SHIFTING TO SURVIVE

Major building projects put on hold were now under way, including additions to the boiler house and main office building, a new laboratory and five additional warehouses. In succeeding years a new laboratory and mill office building, large finished goods warehouse, raw stock warehouse and receiving dock were erected. Old calenders were rebuilt, and new cutters, trimmers, a supercalender and the company's first paper counting machine were purchased. New rewinders to make larger rolls of coated paper also were installed.

The modernization did not end there. "In the old days, to check out dye colors a guy would take a strip of paper and dip it into the tank; if the tint was close he'd say, `That's good enough' and pour it into the coater," says Kneepkens. "Obviously, we needed a more systematic, scientific process."

Lewenstein directed that all tint batches be sent to the technicians for examination prior to their use. "We'd dry the dye and analyze it for color match," Kneepkens recalls. "Eventually, we were being brought other samples to measure coating flow, adherence and other properties." More chemists were needed for the work, and in 1948 Thomas Busch, a recent Lawrence College graduate, joined the technical department.

Other processes also were reworked. Wooden cases to ship papers were replaced with cardboard cartons made by a machine. The power plant was revamped to rely less on steam engines and turbines for electricity. And the primitive festooning practice of drying coated paper on ladders above hot, circulating

Not many companies can advertise their products using the actual products. Appleton Coated's sample kits were made with the actual paper advertised, as in the case of Woodbine Colored Enamel, right, the recognized standard in the United States for colored coated paper. Above, Appleton's research lab in the 1940s. Opposite, Appleton's office employees worked in close quarters, as this photo of the office in 1957 shows.

and checking specialty papers for defects," says Al Kneepkens, hired as a lab technician in 1947. "They'd also count the number of papers, and when they reached 500 — the number for a ream — they'd put a sticker on the bunch."

When the war finally ended, the company experienced a huge surge in orders. "We were besieged," says Bud Heinritz, in charge of scheduling the coaters at the time. "It was a constant battle to meet demand." Profits were reinvested in the company, giving it the financial wherewithal to acquire a substantial swath of adjacent land and buildings from North Meade Street to East Wisconsin Avenue to North Rankin Street. The disparate group of sellers included Holtz Nursery, Northern Transportation Company, Gochnauer Concrete Products, Badger Periodicals, Van Wyk Coal Company and Whispering Pines School. Some buildings were torn down; the rest were renovated and converted.

More than 300 employees worked at the Appleton plant in the early 1950s, a time when the company produced a variety of labeling papers for cans, packages and colored enamel papers, as well as coated one-sided enamel Litho Label paper used for cigarette "cup" packages. The photo above shows the plant from the loading dock side in 1951.

air gave way to more efficient drying in an enclosed tunnel connected to the coater.

After Reeve was appointed plant manager in 1947, he made additional enhancements. "World War II had resulted in the design, manufacture and installation of new handling equipment at the plant, [and] I could see we might gain by [improving] utilization of our space," he said. "So I decided to take a course in materials handling at the University of Wisconsin, [which] was tremendously helpful, and we embarked on an era of improved productivity."

Working conditions also improved. The company installed new lighting and ventilation systems and built more modern locker rooms, lunchrooms and bathrooms. It also introduced its first pension plan in 1947, awarding each full-time, hourly employee who'd served a minimum of 15 consecutive years a pension of "one dollar per month" for every year served. The plan was significantly improved seven years later.

Of all the productivity enhancements, the most important was the replacement of brush coaters with air knife and roll-type coaters. The old brush coaters were unable to provide the dense colors demanded in printing materials. Appleton Coated had bought its first roll-type coater, which used a series of rollers to smooth and level coatings to afford the desired deeper colors, in the early '30s. In 1935 it introduced its first air knife coater, the No. 1 coater, a machine that literally blew excess coating off the paper to assure a level and uniform application.

More coaters were fitted with air knives in succeeding years, including No. 4 in 1937 and No. 3 in 1948. By the early 1950s, brush coaters had become a relic of the past. "Air knife coaters were much faster than either brush or roll-type coaters," says Orv Luebke, who joined the company's coating department in 1945 and operated all three machine types. "They gave us the ability to coat up to 1,000 feet of paper a minute."

Overall, more than $770,000 was spent on the various capital improvements during the decade, compared to $414,000 in all previous years combined. The modernization paid off, and by 1950 Appleton Coated operated 99.5 percent of available man hours, a marked improvement in productivity.

Peak Years for Cigarette Cups

Heinritz's tenure as president came to an abrupt end in 1948. A heavy smoker since his teens, he succumbed to lung cancer after a courageous battle. Charlie reassumed the presidency at the age of 77, returning to lead the company he'd founded four decades earlier. "John Reeve and Dick Mahony were in the best position to succeed my dad, but Mr. Boyd felt they weren't ready," Heinritz explains.

Charlie no longer strolled to work, driving instead in his gleaming black, custom-built Packard Opera Coupe automobile, which he parked within view of his office window. "He'd cross Meade Street in that car at just about the same time the train arrived, barely missing it," Siekman says. "Martha used to worry that her dad would die on the railroad tracks, but he was a stickler for doing everything by routine."

Henritz recalls once seeing Charlie miss a speeding train "by a whisker, and all the while the whistle is blowing. I think his hearing was a bit impaired," Heinritz says. "Maybe that's why he shouted `Ruby!' so loud."

When World War II ended and the government ended production controls over wartime goods, Appleton Coated resumed its prewar manufacturing pace. The volume of paper used by businesses skyrocketed. Bernice Gosz, left, of the sorting department flipped through a stack of sheets representing the amount of paper used by the average American in 1955. New hires in 1954, above, included Marvin Ernst in the electrical department and James A. Stevens in pipefitting.

Paul Truttschel, above, a longtime Appleton sales manager, saw early microencapsulation technology in the late 1940s. He had left Appleton and was working for Georgia Kaolin, a maker of clay whose product was used to coat receipt paper sold by The National Cash Register Company. Truttschel rejoined Appleton and brokered what turned out to the most successful deal in the company's history — a partnership in which Appleton helped NCR devise a way to apply encapsulated coatings to paper.

The lion's share of Appleton Coated business at the time was with American Tobacco Company, "our largest customer by far," Reeve noted. Other important accounts included Ferry Morse Seed Company, which purchased coated paper for its seed packages, and Silvercote, Inc., a maker of home insulation products. The latter company bought reflective insulation paper, marketed as Reflect-O-Ray. "We'd apply aluminum, which has terrific heat-resisting properties, to a brown Kraft sheet of paper and sell it for home insulation," Kneepkens says.

Other specialty products were equally innovative, such as battery separator papers for customers like Union Carbide and Ray-O-Vac. The paper's insulating properties provided higher output and longer life spans for batteries. "We made the battery paper on our smaller, slower coaters because you had to put so much coating on it," Kneepkens recalls.

Appleton Coated sold more than 20,000 tons of paper in 1950, generating more than $6.2 million in revenue and a net profit of $359,000 after taxes. Increased tonnage was only part of the healthy earnings — postwar supply-demand economics had propelled a 60 percent price increase over the past five years.

The good times did not last long, however. Coated papers for cigarette cups, which represented nearly half the company's sales, suddenly became vulnerable to competition from integrated paper mills, which entered the market with a lower-cost product. Attracted by the huge volume the business afforded, companies like Kimberly-Clark, Mead Paper Company, Consolidated Paper Company and S. D. Warren converted their paper machines to produce the relatively simple coatings in long, continuous runs. Once again, Appleton Coated had become a proving ground, working out the problems of a particular coated paper, only to lose the business to the integrated mills when it reached high volumes.

In 1952, the peak year for cigarette cup stock, the company shipped nearly 6,000 tons. Following the loss of the Lucky Strike account, shipments plummeted to 3,500 tons the following year. Volume fell another 800 tons the next year, and by the end of the decade cigarette cup production was a trifling five tons, marking the official end of production. Appleton Coated confronted the loss of its main source of revenue for the second time in its history.

Final Piece of the Puzzle

This time, Charlie would not pilot a turnaround. He had retired as president and chairman in 1952, at the age of 80. Mahony, the company's executive vice president, was appointed to succeed the founder and chart a new strategic course. He would do well to take a page from the founder, who often advised, "Don't look back. Always look ahead."

In early 1953 Mahony called a meeting in his office "to consider suggestions of what we might make to replace this gigantic tonnage," said Paul E. Truttschel, who'd replaced Gene Colvin as sales manager upon Colvin's retirement. The suggestions "did not amount to a hill of beans," he recalled.

"We realized there was no other ordinary, specialty type of paper that would replace the thousands of tons in cigarette paper we were producing," says Sherm Frinak, assistant plant manager at the time. "There was nothing that would fill those carloads every week, certainly not battery paper or insulating paper."

Unbeknownst to the group, Truttschel had the answer to their problems. Although he'd originally joined Appleton Coated as a chemist in the mid-1940s after a short tenure at the Institute of Paper Chemistry, Truttschel left in 1948 for a job at Georgia Kaolin, a supplier of clay. There he was privy to a revolutionary invention by a chemist named Barrett K. Green at The National Cash Register Company (NCR). The Dayton, Ohio-based company had searched since 1939 for an alternative technology to replace ink ribbons inside its cash registers. Receipts printed with ink often were illegible because of users' failure to replace worn ribbons. The company also sought a lower-cost alternative to the sales books it marketed, which used carbon paper interleaves.

While pursuing graduate studies in chemistry at Cornell University in the 1930s, Green was captivated by systems in which a liquid was dispersed within a solid. His knowledge of the subject raised an interesting possibility — a paper that would contain within it all the ingredients for printing. He imagined a type of impact printing paper whereby some kind of mechanism would strike the paper's self-contained "ink" to produce an image. Green earnestly experimented to bring his concept to fruition.

Truttschel saw the process at work when calling upon

International Paper Company, a Georgia Kaolin customer that produced cash register receipt paper for NCR. The company was in the thick of trying to formulate a coating to create impact printing paper. "I was absolutely amazed at the wonders of this development," Truttschel remarked.

After numerous experiments and technical assistance from a roster of big-name companies like Eastman Kodak, Monsanto and DuPont, the project stalled. "It seemed reasonable to assume that, had the development become a commercial success, I would have heard about it," Truttschel said.

The problem was the initial technology, which involved drops of colorless dye solution held within a film coating. The liquid kept leaking and discolored the paper. Green's studies of colloid chemistry, in which one substance is divided into minute particles and dispersed throughout a second substance, proved invaluable in the next phase of research and development. In 1950 he and Lowell Schleicher, a close colleague and methodical NCR scientist, invented *coacervation microencapsulation* — microscopic capsules that could hold liquid dye and oil inside their solid shells. When pressure was applied, the capsules ruptured, releasing the dye. "We finally cracked the thing!" Green exclaimed. "Microencapsulation was the final piece of the jigsaw puzzle."

NCR now searched for a company capable of applying microcapsular coatings to paper.

In a top-secret meeting with Mahony, Truttschel divulged the entire story of NCR's impact printing paper strategy and the extraordinary opportunity it represented for the company that could successfully coat paper with microcapsules. "I told him I knew of no development in paper research [that] held the promise of this new development, [and] if he was willing to put up the money for a research program to assist in bringing this project to a successful conclusion, I was willing to go to Dayton and talk to Barry Green."

With nothing else to replace the lost tonnage from cigarette cups, Mahony had no alternative. Truttschel packed his bags.

When NCR chemists Barry Green, right, and Lowell Schleicher applied for a patent on microencapsulation, the U.S. Patent Office initially balked. "The examiner refused to believe that a capsule existed, that the paper contained nothing more than an oil and water emulsion," Schleicher said. "So I put my equipment and materials on his desk and demonstrated the entire process right in front of him."

A Remarkable Invention

A major milestone in the history of information processing is the development of carbonless paper. Ironically, like many great inventions, this one, too, was predicated on discovering something else.

Barry Green and Lowell Schleicher, the two NCR chemists who share the patent for the invention of microencapsulation, a key component of carbonless paper, were looking to find an alternative for the ink ribbons inside cash registers. Instead they developed a unique impact-printing technology that replaced another messy annoyance — carbon paper.

NCR turned to Appleton Coated to help improve the coating mixture and refine the application. "Mr. Green … appeared to be quite enthusiastic about the possibilities of our doing the job for them," wrote Paul Truttschel in an April 1953 memo to Appleton Coated President Dick Mahony. Not long after, the companies began experimental trials producing carbonless paper, later marketed as the NCR Paper brand.

Previously NCR had worked with International Paper Company to develop a workable microcapsular coating, but the integrated paper mill abandoned the project in the belief that it was unworkable. As of February 1953 NCR still had not found a paper mill or specialty coater capable of producing the product. "We have been to virtually every … coater of consequence in the country, none of whom are able to do the job," Green complained to NCR officials.

Appleton Coated marked the end of the odyssey. "The primary advantage that Appleton had was that it was willing to try our crackpot ideas," says Bob Sandberg, an NCR researcher who'd assisted in the development of microencapsulation. This willingness was driven by the company's dire straits at the time — the loss of its main source of revenue, the manufacture of cigarette cup paper. As Schleicher said about other companies that had failed to come up with an effective coating, "They weren't as hungry as Appleton Coated."

Once production of carbonless paper was under way, major changes were in order at the Appleton plant to accommodate the pressure-sensitive aspects of the product. "The custom at the time was to roll the paper on the floor after coating," recalls Frank Sanders, who started working at the plant in 1950. "Well, we couldn't do that because a tiny pebble would cause the capsules to rupture. … Eventually we put padding over everything."

Although Appleton Coated never had a contract with NCR guaranteeing supply, the company invested heavily to continually improve quality. It was so confident of its product's superiority that it solicited dissatisfied customers to contact NCR Vice President Howard Lauer by telephone with their concerns. Advertisements even listed Lauer's toll-free phone number and the times during the day when he'd be available.

Few customers complained, given the product's advantage over carbon paper, the bane of secretaries everywhere. Many clever company ads in the

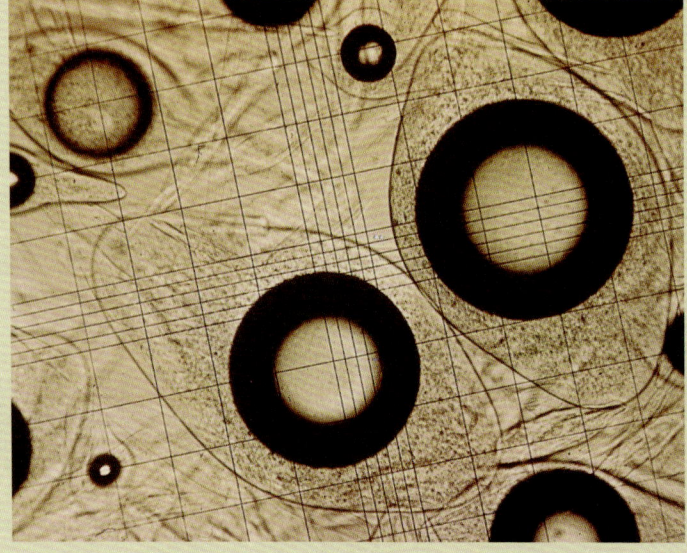

To convey the tiny size of its microcapsules, Appleton often used a photo, opposite, of capsules affixed to a strand of human hair. Right, drops of oil appear as circles inside gelatin capsules in this photograph against a grid taken through a microscope.

1960s and 1970s capitalized on the comparative benefits.

The capsules used to coat carbonless paper came from a plant in Dayton and another built in Portage, Wisconsin, located on a narrow strip of land between the Fox and Wisconsin rivers. It's a famous spot — in 1673, the explorers Father Jacques Marquette and Louis Jolliet portaged there en route to mapping the upper Mississippi River, regarded as one of the most significant events in Wisconsin history. For many decades thereafter, missionaries, trappers, traders and settlers traveling south on the Wisconsin River to the Mississippi and north on the Fox River to the Great Lakes, passed through the Portage Canal connecting the Fox and Wisconsin rivers.

Ralph Montello, plant manager of the Portage capsule plant from its christening in 1967 to 1989, selected the site. "Howard Lauer and I were talking about where would be best to build the plant," Montello recalls. "I suggested Portage because it was relatively close to the University of Wisconsin in Madison, where we could hire graduates in chemical engineering. Barry Green designed the plant."

Following NCR's divestiture of Appleton Papers, the company took ownership of the Portage plant (the Dayton facility closed). Today 10 employees work at the Portage plant, which provides millions of pounds of microcapsular emulsion to the company's three carbonless facilities in Appleton; West Carrollton, Ohio; and Roaring Spring, Pennsylvania.

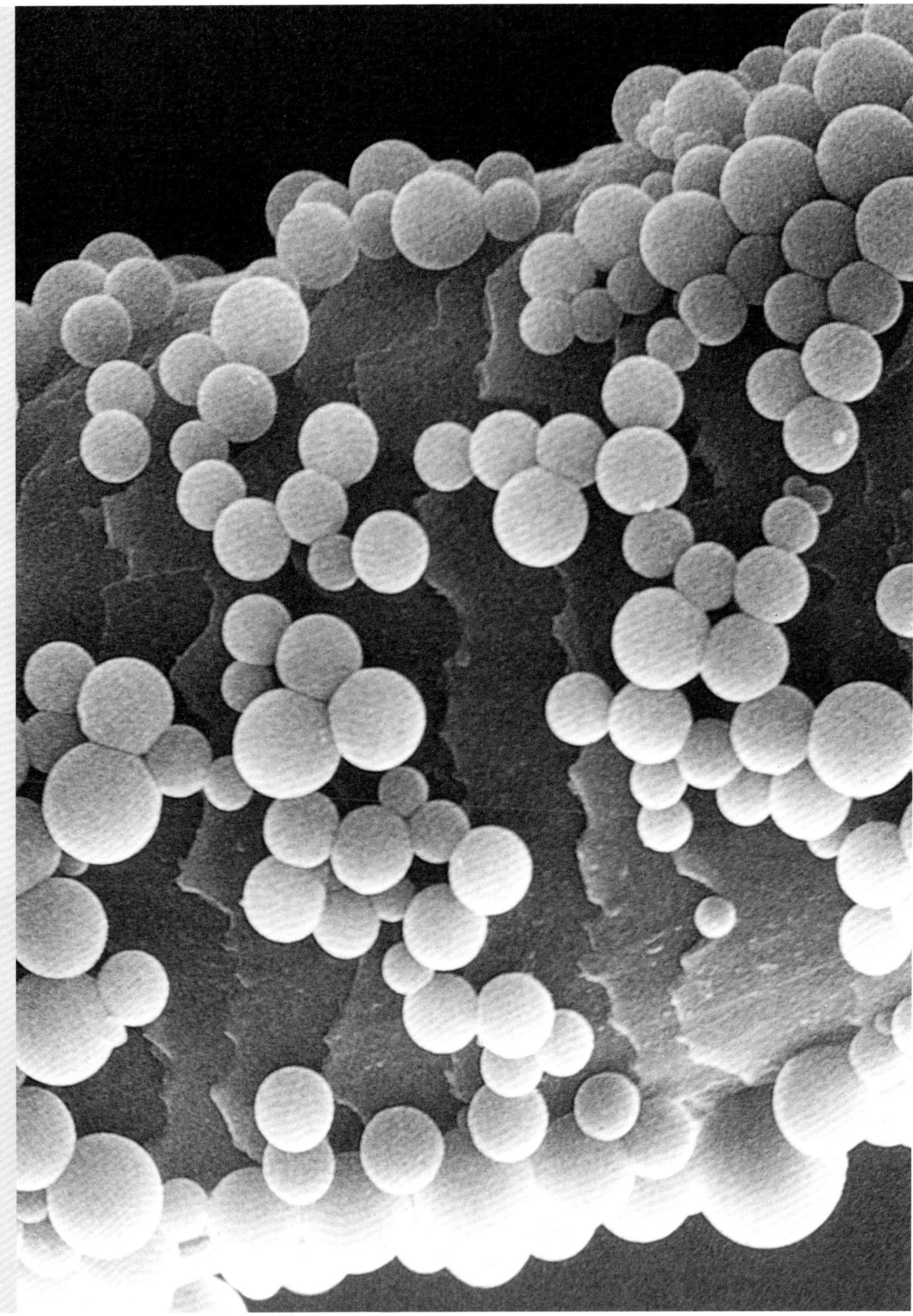

Riding the Carbonless Wave

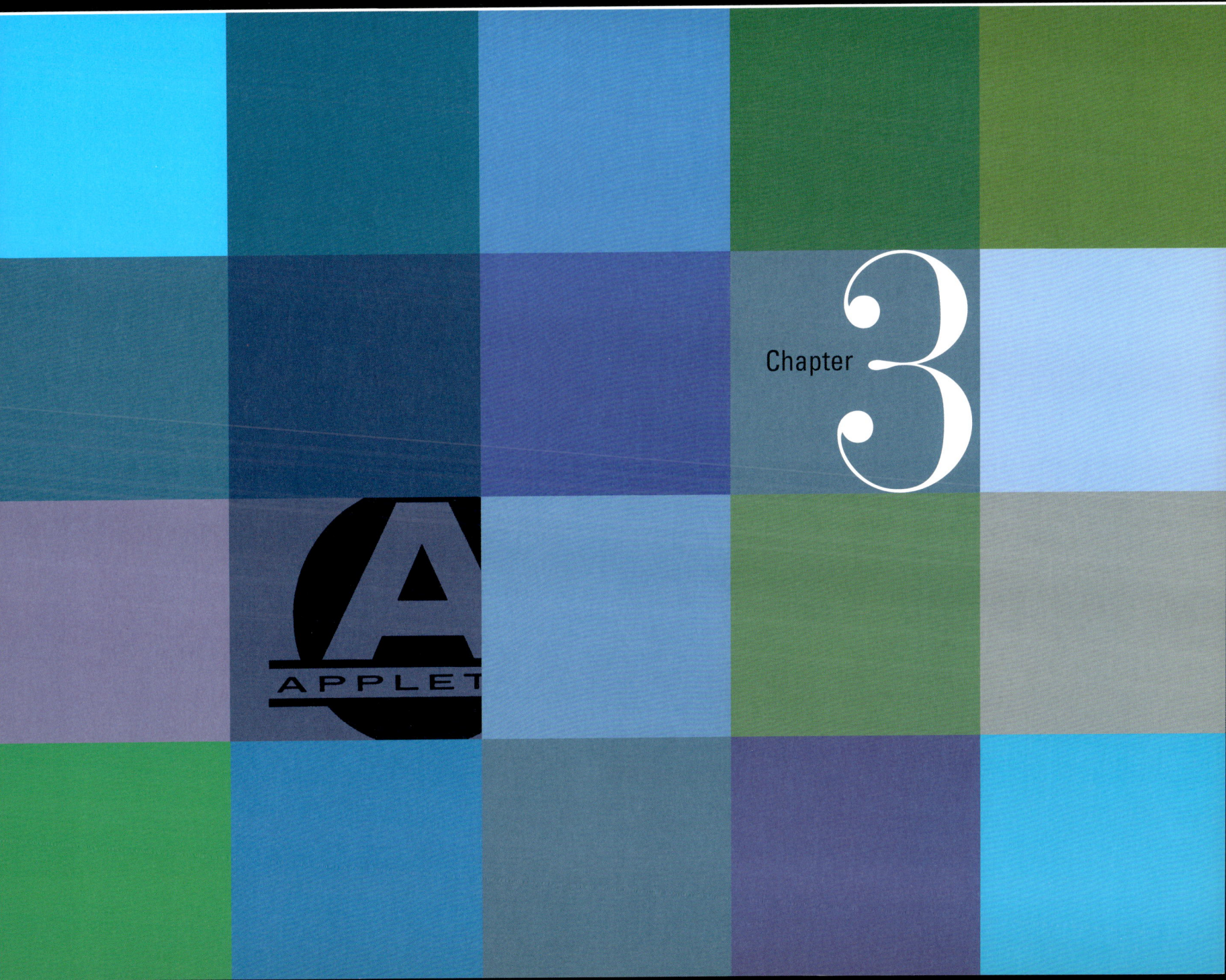

Chapter 3

In an era of technological marvels like Dacron, 3-D movies and television, Appleton Coated manufactured a revolutionary product — carbonless paper, whose birth was recounted, below, in a company newsletter. Credit for the invention of carbonless paper is due to chemists Barry Green and Lowell Schleicher at The National Cash Register Company. Appleton Coated researchers, opposite, left to right, Roland Meyer, George Mickelson, Abe Lewenstein, Tom Busch and Bill Page, were key in helping NCR introduce carbonless paper.

Paul Truttschel got off a plane in Dayton in March 1953 to meet with Barry Green, the co-inventor of microencapsulation. The timing of the meeting could not have been better. Truttschel recalled: "I answered a whole flock of his questions and then Barry told me, 'It is most fortuitous for you to come in and see us just at this time. The R&D is complete, and we are ready to make trial runs with microcapsular coatings.'"

Green was aware of the company's reputation as the most sophisticated paper coater in the world and was impressed by its state-of-the-art air knife coating machines. As he told Truttschel, the equipment "didn't require the application of pressure for coating, which could damage the capsules," he said. "It would seem your equipment is well-suited to the operations required."

Green and Howard Lauer, another NCR researcher, traveled to Appleton the following month armed with a sample of a new microcapsular emulsion in a large jar. They presented the viscous solution for testing to Abe Lewenstein and Tom Busch in the technical department; they were able to dilute and apply it successfully to paper with the No. 1 air knife coater. Satisfied NCR finally had found a coating company with which to undertake trial runs of their new product, Green and Lauer later sat down with Truttschel and Appleton Coated President Dick Mahony. The researchers explained NCR's plans to manufacture and sell a new type of paper competing against carbon paper in the manufacture of multipart business forms, a product they believed would generate $20 million to $50 million per year. Given the perilous position of Appleton Coated following the loss of the cigarette cup business, Mahony's interest was more than piqued.

Charlie Boyd did not live to see these developments. On January 28, 1952, the founder passed away in Miami at 80 years of age. The company closed for business the following day in quiet observance of his profound contributions.

On July 23, 1953, Green again trekked to Appleton, this time with Bob Sandberg, to conduct additional tests. "Barry and I drove a half-ton truck we'd rented, with a roll of paper we'd bought from Mead and a drum of 'open structure' emulsion," Sandberg says. "About the fastest it went was 55 miles per hour, and that was downhill. It took us an awfully long time to get there."

With Green and Sandberg in attendance, Busch conducted additional trials. "We had a miniature, hand-operated … air knife coater in the lab, which utilized hand and foot pedals," he wrote in *Adding Value to Paper*, his incisive 1997 history of Appleton Coated. "It was fortunate that I had musical experience because it took hand-to-foot coordination to operate it."

The tests helped clarify the potential of microencapsulation for use in multipart business forms. To produce the three-part business form without carbon interleaves, three grades of carbonless paper were required, one for the top sheet, another for the middle sheet and the third for the bottom sheet. The top sheet of paper was coated on the back side with millions of microcapsules containing a solution of colorless dyes and oils. Each capsule

The birth of carbonless: NCR Paper* brand

In history there are lessons. What led to carbonless paper's success in the '40s and '50s was a combination of hard work, innovation, quality consciousness, customer satisfaction and teamwork. Those same attributes are vitally important today as we enter a decade promising rapid market growth and its resulting market share opportunities.

The following chronology tracks carbonless paper's development. It is told by people who played major roles in its growth. Their contributions, plus the contributions of countless others, have made Appleton Papers the world's leading producer of carbonless paper.

Barry Green, second from left, consults with, from left, Appleton's Helmut Schwab, material research; Lowell Schleicher, basic; and Don Churchill, exploratory.

Thanks to carbonless paper, secretaries no longer would have to rush to the washroom to clean hands smudged with carbon paper, a fact that NCR capitalized on in its advertisements. NCR even adorned the cover of its 1954 annual report with the illustration above in reference to the breakthrough. Left, a ream of the NCR Paper brand of carbonless paper.

was no bigger than 20 microns (a micron is 39/100,000ths of an inch). This sheet was called CB, for "coated back."

The middle sheet of paper was coated on the front side with casein and a chemically reactive clay receiver material. Upon impact, the colorless dye in the microcapsules on the top sheet changed to an intense greenish-blue color that left their imprint on the front side of the middle sheet. "The dye was crystal violet lactone, and when the capsules broke and the dye came in contact with the acidic surface of the clay, the resulting chemical reaction produced a color imprint," explains Helmut Schwab, an NCR color chemist in the 1950s.

The back side of the middle sheet was identical to the back side of the CB sheet, coated with microcapsules of colorless dye. The coatings on both sides gave the sheet its name —

NCR chemist Barry Green revolutionized the business forms industry with his co-invention of microencapsulation. Green realized that development of carbonless paper required coating expertise that few companies could muster. Appleton Coated made his invention practicable. A promotional carbonless paper sample, below, describes the three-sheet carbonless process, noting its values: "sharp, clear copies, smear and smudge resistant, soil free, saves time and eliminates carbon paper disposal."

RIDING THE CARBONLESS WAVE

CFB, for "coated front and back."

The bottom sheet of the three-part business form was coated on the front side with casein and clay, and was identical to the front side of the CFB sheet. Its name was CF, for "coated front" sheet. In practice, a person or machine that wrote or printed with pressure on the top sheet of the CB paper left a perfect imprint of this marking on the middle CFB sheet and bottom CF sheet.

The ingenious technology eliminated the need for the insertion, extraction and disposal of carbon paper interleaves, saving users time, money and mess — this last point no small matter in use. Carbon interleaves often left their residue on clothing, hands and anything else they touched. The new technology was referred to as "carbonless paper," a name coined by Kircher, Helton and Colett, NCR's advertising agency in Dayton.

The First NCR Paper

One day after Green and Sandberg arrived in their rented truck, Busch made 1,000 pounds of CB on the No. 1 air knife coater at a rate of 350 feet per minute. "The paper ran very well," he reported. "The CB had no apparent visible damage at the coater reel or at the slitter-rewinder." A second machine run was made to produce 1,000 pounds of CFB, also at 350 feet a minute; the results were not as good. "The CFB suffered serious capsule breakage on both center and contact rewinding," Busch noted. "This problem of potential capsule breakage at every step of manufacturing and handling illustrates the magnitude of the task ahead."

The paper was extremely pressure-sensitive, which did not jibe with Appleton Coated's conventional coating processes. "If there was a wrinkle in a roll, the normal practice at Appleton was to hit the roll with a two-by-four to get rid of the wrinkle," Schwab says. "You do that with CFB and a streak will run through half the roll."

Busch's job was to revamp manufacturing and product handling to reflect the pressure-sensitive aspects of the paper. It was not an easy task. "The talk going around Appleton was that, 'you can't make watches in a blacksmith shop,'" he said.

Fortunately, Busch was smart enough not to listen to the doubters. "He was not about to give up," said plant manager

*NCR Paper is a registered trademark licensed to Appleton Papers Inc.

John Reeve. "He kept trying one approach after another until a way was finally found to get the job done."

In early September 1953 the company conducted a second CFB trial on the No. 1 coater using CF paper coated on the machine by two mills — Mead Paper Company in Chillicothe, Ohio, and Combined Locks Paper Company, a producer primarily of telephone book paper, five miles east of Appleton Coated on the Fox River. Although there again was some minor discoloration, NCR was encouraged by the results, and by the end of the year Appleton Coated had produced 100 tons of carbonless paper. NCR subsequently undertook performance tests of the paper with printers and end users, and earmarked funds to build a plant in Dayton to manufacture microcapsules.

On March 26, 1954, NCR Paper — the official trademark — made its market debut. The "NCR" did not stand for National Cash Register, but for "No Carbon Required," a name suggested by NCR's ad agency. NCR made presentations to 50 manufacturers of business forms during the year, but only eight ordered it. Nevertheless, their aggregate purchase of 438 tons "greatly exceeded production," NCR's 1954 annual reported stated.

Many other business forms manufacturers were reluctant to purchase carbonless paper because of their press operators' resistance to the idea. "It was not the standard on their presses, so they had a lot of trouble running it," said Green. "Had customers not demanded the product, printers would never have taken the time to print it."

Jack Wikoff was an NCR sales representative hawking the first sheets and rolls of carbonless paper. "We had seven or eight reps throughout the country in different regional areas that sold it to merchant salesmen, as well as directly to a few

Above are some of the early logos NCR used to brand its "no carbon required" paper. Sales brochures, right and opposite, noted its revolutionary aspects. As advertisements of the period promised, "We offer the solutions because we understand your needs."

RIDING THE CARBONLESS WAVE 61

NCR ADVERTISING

...sensational new

NCR* NO CARBON REQUIRED

Sales and advertising of carbonless paper were overseen by The National Cash Register Company in Dayton, Ohio. The company's advertising agency, Kircher, Helton and Colett, which had coined the term "carbonless paper," was in charge of promoting the new product.

The Dayton-based agency's creative brain trust was in high gear developing advertisements that capitalized primarily on the smudge-free properties of microencapsulated, multi-part business forms. "They can't smear or smudge customer or employee," one ad from the 1960s proclaimed.

Many ads in the 1970s were funny, playfully reflecting the era's changing styles and tastes. Several featured an attractive young blonde in a white miniskirt with knee-high white boots, noting that NCR Paper brand wouldn't mess up her outfit. The "Carbonless Paper for Lovers" ad series, which ran in *Cosmopolitan* magazine with the theme, "NCR Paper. Friend of the Working Girl," showed a man and woman in closeup, with the woman's face smeared with carbon residue.

accounts that were printers," he says. "Early on, the price of the product was a difficult proposition when compared to less expensive carbon paper for business forms. But the end-use advantages of the product soon changed that perspective."

NCR retained exclusive rights to the sale of its NCR Paper brand of carbonless paper, which was handled by its sales department in Dayton. The product was marketed primarily for use in multipart business forms, although legibility problems associated with carbon interleaves made it attractive for other applications, too, including included check duplication, rail tickets and traffic summonses.

Refining the Process

Despite its crucial role in the development of an effective microcapsular coating, Appleton Coated did not have a contract with NCR to produce carbonless paper; nor did it have any promise of future orders. It also wasn't the sole coater of carbonless paper for NCR. Mead, Combined Locks and Nekoosa-Edwards Paper Company made carbonless paper using their own base stock, though not nearly at the tonnage rate produced by Appleton Coated. "All we could do was our level best to provide good quality and dependable service," said Reeve. "We pushed ahead, worked hard and took our chances."

Competition also emerged overseas. NCR negotiated a deal in 1954 with Wiggins Teape, a paper company in England, to make and sell carbonless paper under an exclusive license in Europe. In the mid-1960s it also licensed two companies in Japan, Jujo and Mitsubishi, to manufacture and sell carbonless paper in Asia. "The only safeguard in the business was [our] ability to manufacture the product according to NCR standards, [with] a continuous commitment to improve the product," Busch said.

The company continued to refine the ways in which carbonless paper was coated, cut, handled and shipped. Returns of defective paper were not uncommon, and one time it had to take back a full boxcar load of smudged paper from a Buffalo, New York, customer. Appleton Coated learned from the experience and developed crush-proof shipping cartons. "We all had to work like the devil to design and build new equipment and innovate wherever possible to accommodate the sensitivity of

Carbonless paper was a gift of an advertising agency engagement, ripe with comedic possibilities. NCR's advertising agency had a field day with the NCR Paper brand, as this selection of ads from the 1960s illustrates. The industry-targeted ads ran in a potpourri of magazines and newspapers, including workplace publications *Office* and *Reproduction Review*. The ad with the nurse was featured in *Modern Hospital* and *Medical Economics* magazines, while the ad at bottom right featuring the staid-looking banker ran in *Bank Equipment News* and *Bankers Monthly*.

Appleton celebrated its 50th anniversary in 1957 by publishing *50 Colorful Years*, an illustrated booklet printed on papers coated by the company. The cover was printed letterpress on 10-point gold and white Currency Cover metallic paper. Inside papers included Blue Mellochrome Dull Coated Bristol, 80-pound Tan and Lime Woodbine Duplex Enamel, Orange Mellochrome Dull Coated Bristol and various lithographs on Empress Offset Enamel.

the new product," Reeve said. "New protective packaging, rewinding, sheeting, trimming all had to be worked out. We just took the bit in our teeth and ran with it."

Full-scale manufacturing of carbonless paper began with a shipment of microcapsular coatings from NCR's plant in Dayton. The capsules arrived in truck tankers operated by Kampo Transport, a local Appleton company.

Appleton Coated acquired base paper to make CB from two Wisconsin mills — Bergstrom Paper Company in Neenah and Nekoosa-Edwards Paper Company in Port Edwards, Wisconsin. The company acquired CF paper that had been coated on the machine at Combined Locks and converted it to CFB in Appleton. The final product was then shipped to printers. "Reaching regular production on a commercial scale was the result of … arduous and determined effort, in which the skills of every employee [were] stretched to the utmost," Busch said.

Demand Surges

While carbonless paper consumed most of its attention, Appleton Coated also produced a growing variety of specialty papers in the 1950s. The company funded a survey of printers and merchants by American Color Trends in 1954, which guided it to revise its line of colored papers. Faber Birren, a world-renowned authority on color who'd written numerous books evaluating the ways in which people experience color, was engaged to create the new colors. Birren's *Ink Color Guide* was then sent to distributors, printers and advertising agencies to advise effective color combinations.

Appleton Coated unveiled other specialty papers during the decade, such as Apco Masking Paper, an opaque coated product used by printers to outline photographs or in plate making to block light from non-image areas; Apco Hi-Fold Cover, a less sturdy and less expensive paper than Supertuff; and Suedetone, manufactured for greeting card companies that wanted cards with a soft, suede-like surface. Other novel products included pressure-sensitive label paper for packaged bandages, playing cards, paper wrappers for hot dogs, label paper for the outer cover of record albums and gift wrapping paper. "Anyone who worked here went home with all the paper they ever needed for wrapping gifts," says Frank Sanders, who worked at the plant and later became an Appleton human resources manager.

Some paper products were meant to be discarded after use, such as casting paper for shower curtains. "The paper is coated and printed with the desired pattern, then a liquid plastic is placed on it," *The Post-Crescent* reported. "When the paper is removed, the plastic holds the design."

With demand surging for specialty products and the new carbonless paper line, Appleton Coated needed to expand its plant. The company installed an air knife coating head on the No. 2 coater in 1954. The company purchased warehouses north of Wisconsin Avenue and built pipe and electric shops plus a raw material and raw stock warehouse along the railroad right of way. It also acquired additional property on all four sides of the original plant for parking facilities and future expansion needs, and shortly thereafter enlarged the laboratory and main office building. The decade ended with the construction of a new facility to house the trimmers, some general offices and the ladies-only sorting department.

As Appleton Coated built new facilities, it also modified its machinery to optimize carbonless paper production and quality, as well as to produce more coated papers in large rolls. Previously most production had been in sheets and not rolls, a ratio that would dramatically change in succeeding years. In 1956 Appleton Coated made finished papers available for the first time in rolls of 3,000 yards, packed inside an eight-sided, double-walled, corrugated "drum pack."

Even as buildings, machinery and processes were updated, reminders of the past remained, such as the spittoons that were cleaned out nightly. Workers who did not chew tobacco invariably smoked it, and specific areas throughout the plant were designated as smoking corridors. Each person was assigned a particular corridor and was allowed "a maximum of two smokes in the morning and two smokes in the afternoon," the employee handbook warned. Other rules prohibited the wearing of ties and gloves, "shoes with poor soles" and "clothes too torn or ragged to be safe."

Four hundred forty employees worked for the company in the mid-1950s. As in the past they were a close group, playing on the company-sponsored bowling team in the City Industrial

For many years until 1967, when they were replaced by a computerized sampling machine, women counted and sampled papers in the sorting department by hand, ensuring that the number of sheets added up to a ream (500 sheets) and checking the stock for defects. In the late 1940s, approximately 70 women filled the department, fanning through papers and enjoying each other's company. Kampo Transport, a local Appleton trucking company, delivered microencapsulated emulsion for carbonless paper from Dayton to Appleton. Below, employees in 1954 included, left to right, Chester Westphal, Andy Clark, Ivan Stiltjes, Cliff Hurley, George Regenfuss, Harold Babcock and Bob Schmidt.

Members of the men's softball team at Appleton sit for their official portrait in 1954. Front row, left to right: Al Stage, Roland Glatz, Ed Ferrell, Pat Roche, unidentified. Back row, left to right: Al Kneepkens, Don Coyle, Armin Schabow, Jerry Hiler, Stan Satorius, Don Schulz, Milt Collar.

League or on the separate male and female basketball and softball teams. Golf and ping pong tournaments were other pastimes, and many employees gathered for the annual ice fishing jamboree on Lake Winnebago.

Appleton Coated celebrated its silver anniversary on May 7, 1957, five days after the city of Appleton celebrated its 100th anniversary. More than 5,000 visitors toured the plant to see how paper was coated and to view an exhibition of products. The company hosted two parties at the Butte des Morts country club, one for suppliers and the other for customers, and also sponsored a banquet for the Quarter Century Club, formed in 1946 to recognize employees with 25 years of continuous service. To house all the out-of-town guests, the company rented a dormitory at Lawrence College. Additional festivities included a series of open-house celebrations for employees and their families, as well as area business, educational and civic leaders.

Threats to the Carbonless Business

The 1960s ushered in a momentous decade of extraordinary growth. Total shipments reached a record 31,000 tons in 1960, of which 11,000 tons were carbonless paper. Together they added up to a record $16 million in sales. New business forms companies had entered the marketplace as customers for carbonless paper, and market forecasts indicated that carbonless paper production would grow at three to four times the growth rate of the company's specialty grades combined.

To serve this anticipated demand, work commenced on the largest plant addition to date, built to house new and future coating equipment, a large raw stock storage area and a process quality control center. The two-story wooden house that had once served as the main office was razed to make room for the addition, as was a company garage facing Meade Street. The city of Appleton even granted permission to the company to move the north part of Meade Street 150 feet to the west. The total cost

Members of Appleton's 1953 women's bowling team, showing off their new uniforms, were, left to right, Bernice Gosz, Althea Zurilla, Wanda Rollo, Edna Eggert and Joyce Seidl. Company gardener Quentin DeBruin, above, kept watch on Charlie Boyd's plantings.

Fireside Chats at the Farmhouse

Few companies can boast of operating a historic farm house that serves as a bed and breakfast for its customers, employees and other guests. Following the acquisition of Appleton Coated by NCR in 1971 and its subsequent merger with the Combined Locks Paper Company to form Appleton Papers, Inc., the company has operated the Farm Guest House in the Village of Combined Locks.

The house and the farmland on which it sits were built on the eastern shore of the Fox River by George John Berghuis, who had sailed with his wife Johanna Welhuis Berghuis and their five children from Holland to the United States in 1867. The family settled in the town of Buchanan, later renamed Combined Locks in reference to the nearby locks built to control the powerful current of the Fox River. George is said to have dug out part of the hilly shore to build a lean-to and log cabin, although local legend has it that he built the hill itself as well. The story goes that when 6-foot, 4-inch George came back from a day's work in the fields he'd kick the dirt off his wooden shoes in the same spot. Eventually the dirt piled up and formed the hill.

In 1884 the giant farmer erected a stately brick home, a shed and a barn on the hilltop. Three years later he built a $400 addition to the house and installed a stone marker in one of its gables. Inscribed in Dutch, it reads:

"Oh God Be The Ruler Of This House, And All Who Dwell Therein: G.J. Berghuis and Wife Johanna Welhuis, 1887"

After George died in 1910, his son Barnard inherited the farm. Barney, who is said to have owned the first car in the area, invited two of his sons to farm the 110 acres in the 1920s. They believed tilling the land held no future, however, and sold the farm to Combined Locks Paper Company, which remodeled the farm house into a guest house with private bedrooms and baths. Subsequent additions through the years have provided a meeting room, screened porch, den and other amenities. A large white barn, built in 1926 after the original barn burned to the ground, accommodates summertime business meetings. Recently a modern kitchen was added to the barn, and several bathrooms were upgraded at the main house, which retains its early American flavor. Deer often gather within yards of the porch windows, while hawks soar above, making the executive bed and breakfast a delight for all who visit.

Through the years many NCR executives and their spouses have stayed at the guest house, including former NCR Chairman Robert S. Oelman, according to D. W. "Russ" Russler, an Appleton Papers' finance executive in the 1960s and 1970s. Russler was responsible for approving expenditures to upgrade the facility, a job in which he took particular pleasure. "We had a corporate policy that we couldn't spend more than $25,000 on any capital improvements without final approval from NCR in Dayton," he recalls. "I had to wink a bit because we spent a lot of money on the farm house, but always made sure it was a tad below $25,000. That's how much we loved the place."

It's an Appleton tradition that a tree is planted at the company's Farm Guest House when a chief executive retires. Here outgoing CEO Tom Busch, right, wields the shovel as Executive Vice President Dale Schumaker steadies the tree planted in Busch's honor in 1985.

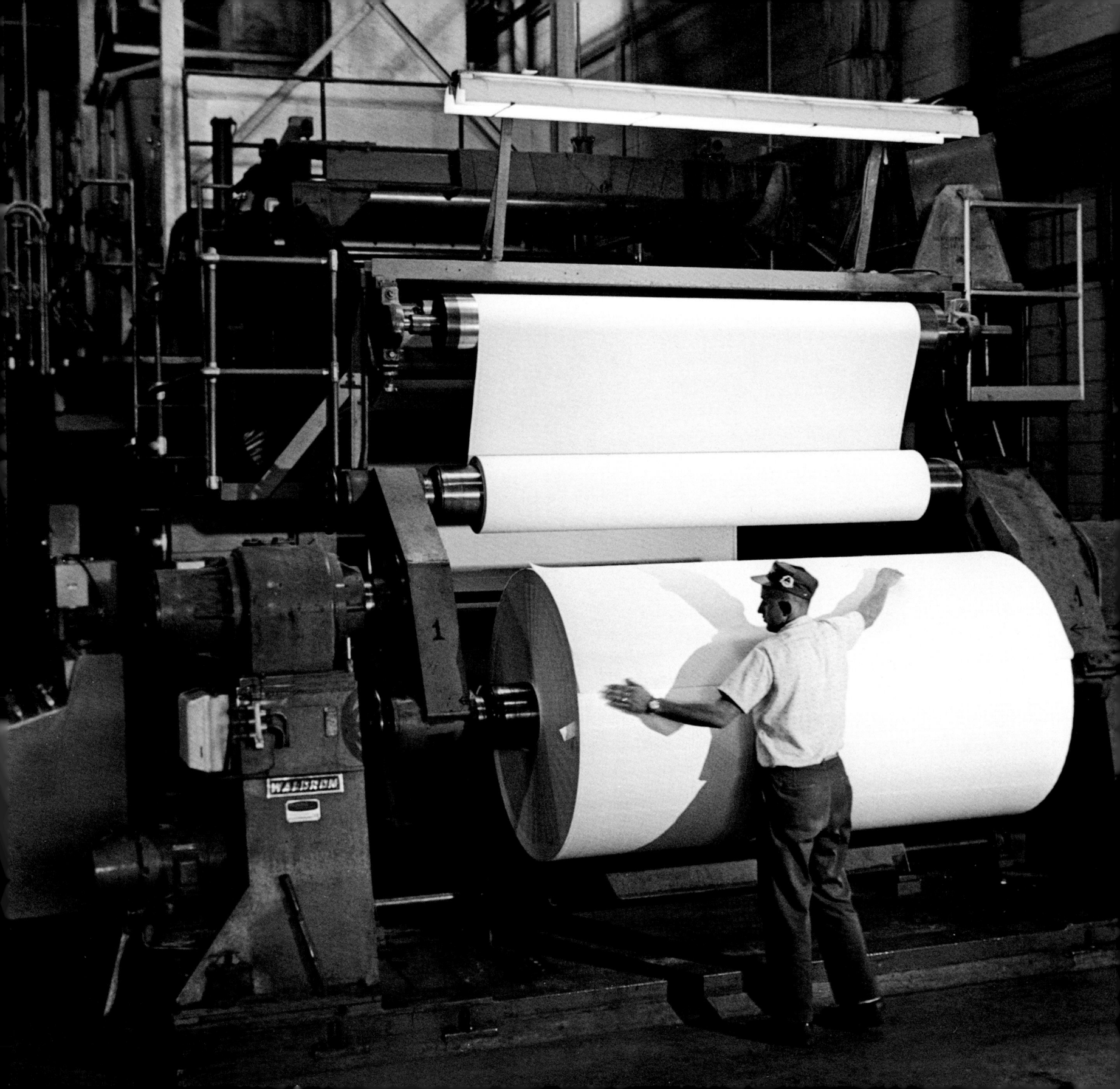

of the plant addition was $5 million — the largest single investment in the company up to that time and one without a guarantee from NCR that it would continue its relationship with Appleton Coated.

A new, high-speed coater (No. 7) was installed in the new building, in addition to related handling and storage equipment. The machine coated paper at 2,000 feet a minute, quite a substantial improvement from the 600-feet-per-minute rate of the company's six other coaters. The coater's production capacity of 30,000 tons per year also was greater than the output of the other coaters combined. By 1961 total manufacturing space at the plant exceeded 430,000 square feet, up from 218,000 square feet in 1940.

After years of harmonious relations, problems emerged between NCR and Appleton Coated. The first work stoppage in the company's 54-year history on August 11, 1961, "caused a major disruption of service to NCR," says Tom Sullivan, who retired as the company's sales administration manager in 1989. "We didn't fulfill NCR's requirements, and they were forced to rely more on Mead, Nekoosa-Edwards and Combined Locks." More than 320 production workers, members of Local 15014, District 50 of the United Mine Workers, had struck over wages and health insurance. Although a compromise was reached in 17 days, the strike reinforced for the company the risk of putting nearly all of its eggs in one basket. More than 60 percent of sales volume and 90 percent of Appleton Coated's operating profit was concentrated with NCR.

Worries mounted when it became apparent that NCR's research division was seeking ways to coat carbonless grades on the paper machine in a single process, a development that would deemphasize Appleton Coated as a supplier since it did not manufacture paper. Both Combined Locks and Nekoosa-Edwards also were gearing up to coat CB and CFB, producing the paper using off-machine coaters located at their mills. Even business forms companies like The Moore Corporation were developing carbonless systems outside of NCR's patents. "We were fully aware of our potential vulnerability without any pulp or paper manufacture, dependence on two paper mills for CF and most of our sales and profit controlled by one customer," said Busch.

These concerns absorbed new Appleton Coated President John Reeve, who succeeded Mahony in October 1962. Reeve had distinguished himself over the years as a competent, trustworthy leader, admired by fellow managers and plant workers alike. "He'd walk through the plant and know everyone, as well as the names of their spouses and kids, and whether or not they had a sore back or a hurt finger," says Sherm Frinak, a former plant manager. "He was a master of communication."

Diversification and New Blood

The new president inaugurated a policy of regular strategic meetings at which he emphasized continual improvements in quality to maintain a consistent demand from NCR. To manage the risk of diminished orders, he urged product diversification. In 1963 Appleton Coated acquired five and one-half acres of adjacent property and buildings owned by Fox Tractor to house an experimental solvent coater to make a wide range of products used in picture frame covers, adhesive-backed papers, book bindings and even the heels on shoes. The water-resistant solvent coatings contained film-forming substances dissolved in volatile organic liquids. The resulting mixture was then applied to Kraft paper or heavier saturated paper.

Another diversification was the coating of heat-sensitive paper, also called thermal paper. In the late 1950s NCR's research division had developed thermo-chromic paper as an outgrowth of its carbonless paper research. A carbonless image develops when pressure is applied to paper coated with microcapsules. With thermal paper, the image is formed by the application of heat; temperature transforms material in the coating from a solid to a liquid, which then reacts with colorless dye, also in the coating, to create an image. "Thermal technology is a kissing cousin to carbonless technology, although one involves capsules while the other doesn't," explains Dale Schumaker, who supervised the solvent coating operation in the 1960s.

NCR began manufacturing thermal paper in 1963. At roughly the same time, 3M Company introduced Thermofax, a machine for making document copies using heat-sensitive paper. The equipment competed against legacy mimeograph

John Reeve, above, became president of Appleton Coated in October 1962 and led the company for 15 years, through diversification, reorganization, a strike and, almost, a sale. Opposite, a plant worker adjusts a roll of coated paper at the reel end of the coater.

Larry Casey, above, was among a number of executives who came to Appleton from other companies in the 1960s to bolster advertising, manufacturing and research. In 1967 the company christened a new research and marketing center to assist its continuing experimentation with microencapsulated emulsions and other paper coatings. Chemist Bill Goetz, opposite, poured from a beaker at the lab in 1968.

machines, new Xerox electro-photosensitive copiers using plain paper and new electro-sensitive machines invented by RCA. By virtue of the company's singular coating expertise (and some adept marketing), Appleton Coated leveraged its experimental coater to coat heat-sensitive paper for NCR's printers and 3M's Thermofax Copiers, as well as electro-sensitive paper for RCA.

The diversification efforts generated modest revenue increases, but nowhere near the volume represented by carbonless paper. To further strengthen the organization, Reeve engaged McKinsey & Company, which counseled the company to restructure finance and operations and to hire additional capable and experienced managers. Reeve took the suggestions to heart. He created a new chief financial executive position and gave the job to D. W. "Russ" Russler, who carried the title of controller. Many other executives came from American Can Company, then part of Marathon Corporation, which had closed its offices in Wisconsin in 1964.

The first American Can executive to join company ranks was Lloyd Swaim as vice president of manufacturing. He was followed by a parade of other managers, including Larry Casey, hired as assistant plant manager and appointed plant manager shortly thereafter; Dale Schumaker, superintendent of the solvent coating operation; and Bob Suess, advertising manager for coated products. Also recruited, though not from American Can, was Ron Jezerc, who joined the technical department. The new blood infused the company with an abundance of scientific, engineering and production talent.

The Restructuring Period

Despite perceived threats to its carbonless business, orders surged through the decade. To accommodate them the company added capacity by installing a second high-speed, electronically controlled coating machine (No. 8) at the main plant in 1965 and constructing a two-story coating preparation building. It also began construction on another major addition on a previously purchased, four-acre parcel to house the No. 9 coater, with room to spare for No. 10. The new building for the coaters, at 120 feet by 405 feet and 60 feet high, was the company's longest, widest and tallest. It boasted a basement-level storage area for millions of pounds of carbonless paper — more than 600 rolls in all, stored three stories high. An operator-occupied cab traversed upwards, downwards and sidewise to individually deposit or retrieve a roll.

The new coaters represented a giant leap in technology. Both were tandem coaters, able to coat both sides of CFB paper in one pass, a revolutionary breakthrough that reduced the company's dependence on Combined Locks and other mills that coated CF paper on the machine. "Tandem coaters changed the way CFB was made and helped us compete more effectively on price and quality," says Schumaker.

Restructuring activities permeated the period. After Busch succeeded Abe Lewenstein as technical director in 1966, he split the technical department in two, with one division focused on specialty coatings and the other on carbonless paper, headed by Bill Page and Ron Jezerc, respectively. The following year, Appleton Coated opened a new research and marketing center in a former Standard Manufacturing Company office and retail building nearby. Like NBC's Today show in New York City, the research laboratory's work was viewable to the public through glass windows.

Swaim, meanwhile, split up manufacturing in 1967, appointing Schumaker to head up coating and Casey to run the finishing operation. Sales were similarly divided, with specialty coated papers separated from carbonless paper, although the latter really wasn't a sales effort per se, since all orders came from NCR. Howard Lauer handled NCR's carbonless business in Dayton while Paul Truttschel handled sales and marketing of specialty coated papers at Appleton.

The specialty coated papers broke down into two product lines — printing and decorative papers like Woodbine Duplex Enamel, Supertuff and Currency Cover; and technical papers such as battery separator paper, electro-sensitive, metallic and thermal papers. There were 26 different kinds of printing and decorative papers and nearly as many technical papers. The latter comprised papers used for clock dials, blotter lining, record labels, tire wrap, wallpaper and window shades. More than 40 mills in 15 states supplied 55 grades of paper to the company in a variety of weights.

Its bustling business required a workforce of more than 720 employees in 1967 — a payroll and benefits tab that added up

A new research and marketing center was erected the following year on the site of a former lumberyard that had been torn down and turned into a parking lot. Like NBC's *Today* show in New York City, the research laboratory's work was viewable to the public through glass windows.

The Roaring Spring Mill

Built in 1865 and put into operation in April 1866, the Roaring Spring mill, one of two papermaking facilities owned by Appleton, is one of the oldest continuously run paper mills in the United States, exclusive of the occasional shutdown. Although the mill's original name is undocumented, we know that it produced paper for grocery bags from pulp stewed with straw and gunnysacks.

The mill today occupies 332 acres in the rolling forest land of the Allegheny Mountains in south-central Pennsylvania, roughly 100 miles east of Pittsburgh. The area is blessed with an ample supply of coal and an active natural spring that continues to supply area residents' water needs. It was the abundance of these natural resources that encouraged John Eby, D. M. Bare and John Morrison to locate their paper mill at a hamlet of eight or nine houses then known as Spangs Mills. Their objective was to build "the pioneer paper mill of central Pennsylvania," Bare wrote. Spangs Mills later became Roaring Spring, a name touting the powerful sound emitted by the spring.

After the mill opened, the owners rapidly encountered a series of setbacks, including financial troubles and a shutdown of several months' duration following a boiler explosion. Before the mill's first year ended, they suffered a more devastating blow when the entire mill burned to the ground. Few believed it would be rebuilt.

Yet by 1867 the mill again was up and running. Rags replaced gunnysacks in the pulp brew in 1870 and, in subsequent years, the mill produced finer, supercalendered grades like book, envelope and mimeograph papers. A second papermaking machine was

installed in 1873, giving the mill a production capacity of 1,860 pounds of paper a day, at an average selling price of 10 cents per pound. A third machine was installed in 1898. By then the mill was listed as part of D. M. Bare & Company. It was incorporated under the laws of Pennsylvania in 1907, and everyone in the area referred to it as the Bare Mill. Daily production grew to 51,000 pounds, although prices plummeted to $3.27 per pound.

Bare died in April 1925 at the age of 91. Combined Locks Paper Company purchased D. M. Bare & Company in 1946, some time after the pulp mill had been converted from a soda process into a Kraft process including a then-revolutionary boiler, considered the best of its kind.

On May 24, 1951, the mill suffered another disastrous setback. The digester room exploded, destroying the pulp mill and damaging many adjacent facilities. Two employees were killed, and it was considered a miracle that more did not lose their lives. The blast could be heard five miles away. Newspaper accounts quoted area residents who'd thought an atom bomb had been dropped, not a far-fetched notion during the Cold War years. "There was some doubt whether the mill would resume," said Vance Myers, who retired from the mill in 1980.

Once again the mill defied expectations, resuming operations in 1954. The pulp and papermaking facilities were renovated, more workers were hired and the economy of Roaring Spring, which depended on the company, revived. The mill's daily production rate soared to 100 tons, more than double the rate prior to the disaster.

Until the carbonless paper era changed the mill's focus, its main grades were tablet paper for notebooks and other school supplies, offset paper for making maps and foil liner paper used to keep cigarettes fresh in their packages. "We also made the base stock for those little blue stamps you see on the tops of cigarette packs," says Keith Paden, who began working at the mill in 1955.

Today the Roaring Spring mill produces carbonless paper, and it has been transformed into the focus of Appleton's security paper production. While the native spring that gave the town its name no longer roars due to changes made in its course through the years, it is a perpetual reminder of the area's bounty and the role it played in building a resilient, "pioneer paper mill."

Appleton's Roaring Spring mill converts trees into base paper that is coated on a paper machine. First, logs are stripped in a debarker, above. Later, left, pulp made from those logs is distributed across the wire on a paper machine to form a continuous web; that web is later coated and wound at the mill to produce 5,000-pound rolls of coated paper.

to $5.7 million. The Appleton complex boasted 750,000 square feet of space spread across 18 acres. A variety of new equipment was on the plant floor, including a computerized inventory control system and a statistical sampling machine in the finishing operation. The latter eliminated the per-sheet inspections done by the ladies' sorting department, but most of the women found jobs elsewhere in the plant. "Lloyd [Swaim] and I had decided that whatever we did to eliminate jobs would not result in layoffs," Casey says.

The plant had also invested in new "cut-sized" sheeting equipment to improve productivity. "Previously, everything that came off the sheeters was slightly oversized and had to go to the trimmers to be shorn to the exact size," Casey explains. "Dale [Schumaker] and I decided to buy new sheeting machinery that cut the paper to size, thus eliminating the separate trimming operation."

Even legacy materials underwent change, exemplified by the use of resin-based CF coating, which provided smoother printing surfaces for greater permanency.

In 1967 NCR built a small plant to make microcapsules in Portage, Wisconsin, halfway between the plants of Nekoosa-Edwards and Appleton Coated. The company initially was anxious that Nekoosa-Edwards would start producing CB for NCR, although this did not come to pass. In future years the Dayton plant would produce an array of products using microencapsulation.

A Strike and Thoughts of a Sale
John Reeve added "chief executive officer" to his title in 1966. The following year, Bill Siekman, Charlie Boyd's son-in-law and the company's industrial relations director, succeeded Mahony as chairman of the board. Mahony was the last link to the past, having begun his career in the cost department in 1917 and risen to the presidency in 1953. He had steered Appleton Coated into the production of carbonless paper at a time when other coating competitors, such as S. D. Warren, elected not to enter the market and forever rued it.

Reeve called the shots while Siekman looked after shareholder interests. The CEO met a stiff challenge on August 1, 1968, when 450 workers staged a walkout over issues concerning a recent contract offer. Starkly reminded of the slowdown that occurred during the 17-day strike in 1961, Reeve was determined to meet as much of NCR's demands as possible. "When the union struck, NCR immediately turned to Mead for more production, and Mead turned on all their capacity," Casey says.

During the strike, salaried workers wore two hats — their regular jobs and new positions on the plant floor. Casey, for example, operated the roll stacker system, while Jezerc and Schumaker ran the rewinders. Even Reeve pitched in and drove a forklift. "We worked 12 hours a day and through the night sometimes, with management meetings on top of that," says Schumaker.

In the middle of the strike, "Reeve made a visit to NCR in Dayton and was told we were a non-dependable supplier and they would replace our capacity," Schumaker adds. "At the time, we'd been thinking about putting in the No. 10 tandem coater, which now seemed to be off the drawing board. Not for John, though. Despite NCR's threats, he turned to our chief engineer, Art Rouman, and said, 'Art, how long will it take to install that damn coater, because we just bought it!' What incredible brass."

During the five-and-one-half-week strike, the plant ran at between 30 percent and 50 percent capacity. When union workers finally returned to their jobs, they were dissatisfied with the settlement. "Money-wise it was a wash," recalls Mickey Thompson, who worked in the sheeter department at the time. "We felt the UAW did a terrible job negotiating for us, so after the strike we voted them out and joined the United Paperworkers International Union." The 1968 strike was the last by organized labor at the Appleton plant. Union employees of Appleton today are members of the United Steel Workers.

After years of nail biting over the competitive dangers lurking in the carbonless paper business, Appleton Coated realized its worst fears in 1969. NCR informed the company that it had purchased The Combined Paper Mills, parent company of Combined Locks Paper Company and the D. M. Bare Paper Company in Roaring Spring, Pennsylvania. The $32 million stock transaction was reached without prior discussion with Appleton Coated management or any other suppliers of carbonless paper. In effect, NCR now placed itself in direct competition with its suppliers.

John Reeve, left, revered as a "people person" who treated everyone the same regardless of his or her position, started with Appleton in 1934 and rose through the ranks from sales assistant to plant manager to president. A plaque, above, mounted on the company's new Wellness Center in 1994 dedicated the facility to Reeve, who is a member of the Paper Industry International Hall of Fame.

Shaken, Appleton Coated's shareholders began entertaining thoughts of selling the company. "After NCR bought Combined Mills, the Siekmans were scared half to death," says Russler. "They were well aware that Appleton Coated was not a low-cost provider of carbonless since it didn't make paper."

Meanwhile, the company's value was at a high point — it had achieved sales of $47 million in 1969, of which $36 million came from carbonless paper. Altogether NCR sold approximately 50,000 tons of carbonless paper during the year, 60 percent of it manufactured by Appleton Coated. "The timing was right for a sale," Russler says.

Without confiding in NCR, Appleton Coated began secret negotiations with a parade of paper companies interested in plunking down their pennies, including Kimberly-Clark, Nekoosa-Edwards, Potlatch, Weyerhaeuser, Champion, Consolidated, Boise Cascade, Olin and Glatfelter, among others. Even non-paper firms like the tobacco conglomerate Philip Morris expressed interest in acquiring the company, as did a group of board directors that included Reeve. However, the Siekmans valued the company "at $30 million to $34 million," according to Busch, and that was more than the directors could afford. "We were leaning toward the Philip Morris offer, which was nearly ten times what the inside group had offered," Siekman confirms.

Having been pushed out of the cigarette package business more than a decade before, Appleton Coated ironically was on the verge of joining a major tobacco conglomerate. Then a surprising suitor entered the picture.

and Head it North"

Chapter 4

The 1970s marked a dramatic change at Appleton Coated. The company was merged into its chief customer NCR, then the world's second largest maker of business equipment. Making the decision was the company's board of directors, shown opposite, from left to right: Tom Busch, Richard Mahony, John Reeve, Martha Boyd Siekman, D. W. Russler, Arthur Allyn, Bill Siekman, William Buchanan and Leonard Haddad. Bill Siekman, above, is a former Appleton chairman and the son-in-law of Appleton founder Charlie Boyd.

Philip Morris fit the mold for what Martha Boyd Siekman, by far Appleton's largest shareholder, had in mind as an acquirer — a buyer that would protect the company's independence, assure continuity in management and employment and keep the company in the community.

Negotiations progressed behind the scenes, given concerns about how NCR would react to a prospective sale. Fissures in the companies' relationship had developed: the Moore Corporation, the world's largest producer of business forms, contracted Appleton Coated to coat its private brand of carbonless paper, which was immune to NCR patents. While the business would reduce the company's dependence on NCR, NCR expressed apprehension about Appleton Coated using its processing expertise to assist a competitor, particularly one that was one of its top five customers.

The negotiations with Philip Morris came to light when the tobacco conglomerate requested that Appleton Coated sign a supply contract with NCR. For 16 years NCR had dealt with its suppliers on a non-contract basis and it was not about to change its stance. Rebuffed, Philip Morris pulled the plug on further negotiations. At the same time, Moore changed its mind and decided to install its own coaters to develop carbonless paper.

Appleton Coated confronted the dismal prospect that former Appleton CEO Tom Busch described as being "reduced to a very small company of low-volume specialty grades and low-margin printing papers, without carbonless or thermal paper."

The board of directors examined the unfortunate turn of events at length. In discussing another potential acquirer, board member William E. "Bill" Buchanan, whose family owned the Fox River Paper Company in Appleton, made an intriguing suggestion. Buchanan was a personal friend of NCR Chairman Robert S. Oelman, with whom he'd attended Dartmouth College in the 1930s. If the board approved, Buchanan agreed to contact Oelman to determine NCR's interest in acquiring Appleton Coated. "Bill commented that we were doing so much work for NCR that we should look to them as a possible suitor," says Bill Siekman.

The former classmates met and chewed over the tantalizing opportunity. In May 1970, eight months after its acquisition of Combined Mills, NCR submitted a proposal for merging Appleton Coated into NCR. The following month, the boards of both companies agreed on a non-taxable stock exchange to effect the merger. At the time, Appleton Coated had close to 900 employees, making it the largest employer in the city of Appleton, while NCR was the world's second largest producer of business equipment, having grown from a one-room workshop in 1884 to a headquarters complex of 33 buildings in Dayton.

NCR now owned two paper mills, in Combined Locks, Wisconsin, and Roaring Spring, Pennsylvania; two microcapsule manufacturing plants in Dayton and Portage, Wisconsin; and the world's most respected paper-coating operation in Appleton Coated. Added up, NCR had the capacity to produce 350,000 tons of carbonless paper a year.

On June 21, 1971, NCR made the decision to unite Appleton Coated and Combined Paper Mills into a single subsidiary known as Appleton Papers, Inc. and headquartered in Appleton. The internal merger would improve coordination between the subsidiaries, NCR stated, and also provide financial benefit. "Appleton Coated was profitable, and Combined Locks had a huge tax loss carry-forward," explains Russ Russler. "Combining the subsidiaries permitted us to use Locks' tax loss carry-forward, reducing our overall tax bite."

John Reeve was named president and chief executive officer of Appleton Papers and was also given oversight of the two capsule plants. Five vice presidents fell below Reeve on the organizational chart: Howard Lauer from NCR, who headed the carbonless paper division; Paul Truttschel, vice president of marketing for commercial papers (Appleton's legacy specialty papers); Lloyd Swaim, vice president of manufacturing; Busch, vice president of technical research and development; and Russler, vice president of administration and finance.

The senior managers were complemented by a group of

Appleton spent $35 million in the mid-1990s to expand the Appleton plant, including the installation of the No. 17 thermal coater, far right. Rolls of stacked paper, right, await retrieval from the plant's automated storage racks for coating.

about 40 NCR sales, marketing and technical personnel, who relocated to Appleton with their families. "NCR was having financial difficulties and decided to abolish research in Dayton, which was about a 200-person department at the time," says Paul Phillips, an NCR chemist who made the migration in early 1973. "Tom Busch arranged for a group of about 15 of us to be transferred to Appleton's research organization, including Lowell Schleicher, who would head research and would report to Tom." Both men had worked closely together on the early carbonless trials.

Research at the time was divided into two areas — basic and applied. The latter was guided by Ron Jezerc, who also was in charge of quality assurance. "Several NCR researchers joined my team, including a few dye and resin developers," Jezerc recalls. "Suddenly, we had all this intellectual capital in Appleton, not just technical research but also sales and marketing. No longer did Dayton run the carbonless business."

As former plant manager Sherm Frinak put it, "The brains of the whole operation were here."

More Uses for Capsules

The acquisition caused some tribulations in the marketplace. Mead Paper Company, a major supplier of carbonless paper to NCR, alleged that the deal abused antitrust laws. "Mead

Although carbonless paper remained its primary product focus, the company continued to make several legacy coated papers in the 1970s, including paper for batteries, such as used in the flashlight below, and Ascot, right, a coated plastic sheet that looked like paper but resisted tearing. The coating was applied to Tyvek, a material developed by DuPont, for use in maps, wall coverings and aisle runners. These and other innovative products are attributable to the brainpower of Appleton research and development scientists.

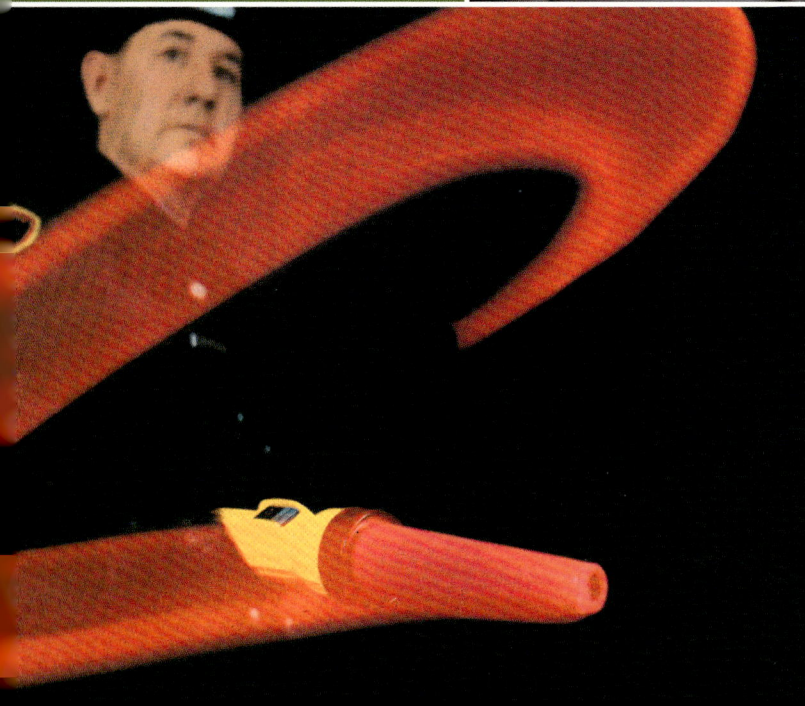

argued that it had invested all this money in assets to produce carbonless paper, and now that NCR owned both Combined Mills and Appleton Coated it would cut them out of the game," explains Dan McIntosh, a member of NCR's technical services group who made the move to Appleton Coated. "To assuage Mead's concerns, NCR gave the company the right to operate as a producer of carbonless paper under its existing patents for free, in effect allowing Mead to make and sell its own line of carbonless paper." Mead accepted the olive branch and launched a competing line called TransRite.

A more difficult burden was consolidating the various plants into a smoothly running whole. In this effort, the most pressing need was transforming the three paper machines at Roaring Spring and the four at Combined Locks to coat CB and CFB carbonless grades on the paper machine — not an easy task given the product's pressure-sensitive characteristics. Roaring Spring also had never coated paper; its competency was producing bond and tablet paper and foil liner for R. J. Reynolds. "Doing CF coating on the paper machine was pretty simple, but we wanted to also do the capsule coated grades on the machine, which hadn't been done before," says Jezerc, who was tasked with this responsibility.

Jezerc traveled to Sweden with fellow researchers and engineers to examine a new type of coating equipment that seemed to fit the bill. The company purchased and modified the coater at Roaring Spring and attached it to the No. 3 paper machine. "We created the world's first tandem coater on the paper machine for CFB production," Jezerc says. "This was an important breakthrough that really enhanced productivity."

Once all three plants were manufacturing carbonless grades, Larry Casey coordinated aggregate production to feed customer orders. "My job was to plan manufacturing capacity and assign the incoming orders to the various plants," Casey says. "This was seat-of-the-pants at first, but we soon put together some spreadsheets and computerized models to give us the most production at the least cost."

The capsular division under the supervision of Jim Herbig didn't only produce microcapsules for coating carbonless paper. It endeavored to develop capsular products serving unique needs in the pharmaceutical, industrial and consumer markets. For example, encapsulation masked the unpleasant taste of some drugs and delayed a drug's release in the body, as was the case with time-released aspirins. Encapsulated mint and menthol scents were used to give flavor to cigarettes, and encapsulated shoe polish was used by manufacturers to release polish gradually through the application of pressure.

In the 1970s the plants developed encapsulated liquid crystals that reacted to different temperatures by producing color. Functional uses included digital thermometers and tumor-detection equipment. The encapsulated crystals also were used to make mood rings, a popular jewelry item of the era. Ring manufacturers claimed that the color changes in the ring were attributable to the wearer's mood and that the ideal color was turquoise, which reflected tranquility. Scientifically, of course, it was temperature and not the wearer's mood that changed the ring's color.

So-called scratch 'n' sniff applications also used encapsulated crystals. "Rub the paper and inhale scents from bacon to bayberry, from garlic to gin, and from sardines to soap," Appleton advertised. "The nose smells what the eye sees." Indeed, no idea was too far-fetched for researchers, who unveiled encapsulated coffee aroma in instant coffee, encapsulated perfume on cash register receipts, and even wallpaper coated with microcapsules — an idea that was revisited in 2002. "We figured you could use a roller to press what looked like plain paper on the wall and instantly have wallpaper," says Siekman, who witnessed a demonstration. "We could never get the capsules to break when we needed them to."

Other non-paper products were equally novel, such as transparent overhead projection film marketed as Teacher's PET, for Projectable Economical Transparency. Users could write on the film with colored markers, felt-tip pens and grease pencils and then project the image on a screen. Then there was Ascot, a coated plastic sheet that looked, felt and folded like paper but resisted tearing. In making Ascot, the company applied specially formulated coatings to Tyvek, a material developed by DuPont, to provide a high-fidelity printing surface.

Ascot offered superior strength, durability and resistance to moisture, rotting, scuffing, curling and ultraviolet light. These attributes made it a practicable product for roll-down maps,

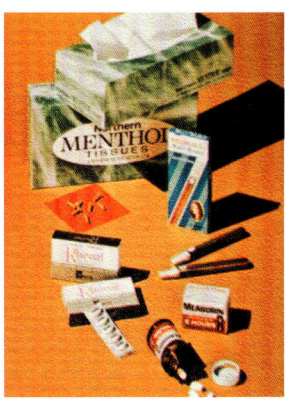

Carbonless paper wasn't the only application for microencapsulated emulsions, as the array of products above indicates. In the 1970s temperature-sensitive encapsulated liquid crystals were used in digital thermometers, tumor-detection equipment and mood rings — jewelry whose color purportedly changed to reflect a person's frame of mind. Scratch 'n' sniff applications also used encapsulated crystals, which were similarly applied to tissues to impart a pleasing scent.

A worker at Appleton's Portage plant, above, mixes microcapsular emulsion for carbonless paper. Opposite is the No. 4 coater at the Harrisburg plant.

wall coverings and aisle runners. Other applications included outdoor signs, placemats, recipe books and sterile packaging for disposable medical devices. "We did especially well selling Ascot to publishers of school books as a book cover," says Jerry Thiel, who joined the company in 1957 as a sales representative. Today Tyvek is more widely known for its use as a vapor barrier in building construction.

Many new products were the brainchild of Ed Mendels, the company's business development manager who'd coined the name Ascot, an acronym for Appleton Superior Coated Tyvek. Mendels also developed Eve, a brand of translucent printing film used in catalogs, annual reports and direct mail advertising. Innovation wasn't confined to non-paper products. When it was clear that some document copiers in the marketplace were "blue blind," unable to copy the greenish-blue images in carbonless business forms, researchers created a modified colorless dye, calling it the Deep Blue System.

Another step forward was in the microcapsules containing the dye, which customarily were manufactured from gelatin and gum Arabic. "Gum Arabic was a natural product from the Acacia tree used in foodstuffs. Prices were competitive," says Jezerc. "It also came out of Africa, where it was susceptible to droughts squeezing the capacity. We worked hard to develop an alternative using synthetic materials that were sourced domestically."

A new synthetic capsule was developed that cost less and provided a superior coating. "You could coat the paper at much higher solids than you could with gelatin, which meant there was less water to dry off," says Paul Phillips, a former NCR researcher who was involved in the research. "Quicker drying translates into higher productivity."

Other refinements in manufacturing were predicated on growing environmental awareness and sensitivity. In 1971 the company eliminated the use of polychlorinated biphenyls (PCBs) in its manufacture of carbonless paper. PCBs were used by companies in a range of industrial applications including electrical equipment, transformers, hydraulic fluid, solvents, adhesives, paints and printing inks. Public concerns about the persistence of PCBs in the environment and their possible impact on people and wildlife caused the federal government to ban the manufacture and use of PCBs in 1978.

Earlier in the decade, the federal government took action to promote workplace safety by creating the Occupational Safety and Health Administration in 1970. "I was appointed safety and training supervisor to coordinate our efforts to meet OSHA compliance," says former plant manager Tom Cashman. "We managed to accumulate 1.9 million safe hours in a row and were three weeks away from hitting two million safe hours. We managed 1,946,322 hours altogether, and I was so proud of it that I bought everyone a sweatshirt with that number printed on the front."

Booming Innovation

Shipments of carbonless paper continued their upward trend through the 1970s. Advertisements touted the smudge-free properties of the paper, as well as its environmental benefits. "Switch a million 4-part forms to NCR Paper and save 30 trees," one ad proclaimed.

To provide faster and more cost-effective service to the rapidly growing East Coast carbonless paper markets, the company built a new plant in 1973. On a 23-acre site five miles southwest of Harrisburg, Pennsylvania, Appleton constructed a new converting plant, which received base paper from the Roaring Spring mill. The 125,000 square-foot plant initially housed rewinding and sheeting equipment for carbonless paper. It further served as an important warehousing and distribution facility, given its prime location near the Reading Railroad and Pennsylvania Turnpike.

Two years hence a 153,000 square-foot finishing, packaging and storage facility was added to the plant to process nearly 100,000 tons of carbonless paper made on the paper machine in Roaring Spring annually. "CB paper coated on the machine in Roaring Spring was sent to Harrisburg for finishing," explains Casey, director of central planning at the time. "Once we converted another machine in Roaring Spring to provide CB, this too was dispatched to Harrisburg."

Soaring demand for carbonless paper required extensive rebuilding of the No. 1 coater at Combined Locks and the purchase of another high-speed tandem coater (No. 12) for the Appleton plant in 1976. The next-generation coater had the

The cover of a sample book points out the advantages of carbonless paper versus carbon paper for use in multipart business forms, the primary market for the NCR Paper brand. By 1977 total sales of the brand exceeded one million tons.

potential to coat carbonless paper at twice the speed and output of previous coaters, thanks in part to the new synthetic capsules, which provided much faster drying. "We could run the machine at 3,000 feet a minute with increased product quality," says Bill Goetz, a manager in the process engineering lab at the time.

Busch and the technical research team had developed the new coater. "Tom put together the new capsules with all this new drying equipment, and it was a gold mine," says Dale Schumaker. "The only problem was the machine cost twice as much as it should — $5 million, as opposed to $2.5 million. Nevertheless, it took no time at all for it to provide a solid investment return. After that it was a money machine."

The decade boasted many other successful research efforts, such as the introduction of microencapsulated colorless dye that left a black image. "There was a desire among business forms makers for a black print similar to the color produced by carbon interleaves, since black ink copied better," Jezerc explains. The patent for the new dye, called N102, was awarded to company researcher Charlie Lin.

The business forms industry also drove paper mills to lower the basis weight of paper to allow more sheets to be used in their forms. "Our mills began to make lighter-weight products, which worked against our tonnage objectives and were much harder to coat but nonetheless satisfied the demand," says Tom Sullivan.

Innovation was so pronounced during the period that patent applications literally flew under the door of Paul Phillips, who had passed the patent bar exam and now headed the company's patent office. "I received at least one application a month from research and engineering," he says. "It was extraordinary."

As in the past, carbonless sheets were distributed through merchants, who sold them to business forms makers and printers; rolls were sold both through merchants and directly to printers. The difference was that Appleton Papers now controlled sales and marketing.

Among its larger merchant accounts were Hall Paper, Copco, Diem & Wing, Zellerbach and Clampitt. "We'd put on seminars to help the merchants sell carbonless, but after our magazine advertising, the product essentially sold itself," says Herb Hofferberth, Appleton Papers sales representative in the 1970s.

Carbonless paper sold so well that plant workers "could dictate how much money we wanted to make," says Mickey Thompson, now the local union president. "We couldn't get enough product out the door to fill the orders. We'd work six days one week and seven days the next. Some guys worked six months without a day off." To alleviate the job burden, the union negotiated a swing shift for workers in 1973.

Barely making a dent in the sales revenue was the loss of carbonless and thermal licensing royalties from NCR patents, which expired during the decade. "This was good income — a half million to two million dollars a year," says Schumaker. "But the real loss was the opportunity to confer with licensees like Jujo in Japan. Through the years we'd learn about new pieces of equipment the Japanese were developing in the thermal area, for which we'd develop paper." Such was the case with thermal chart and calculator paper, introduced by Appleton Papers in 1973.

The unique clothing style and color of the 1970s, above, were a boon for advertising the smudge-free properties of carbonless paper. At left, an engineering marvel that measures one city block long and 45 feet high, Appleton's roll-stack system for storing coated paper uses an automated crane to retrieve 5,000-pound rolls of paper.

APPLETON LOGOS

APPLETON
What Ideas Can Do®

As The Appleton Coated Paper Company evolved over 100 years, it altered its name and logo several times to reflect its primary areas of concentration, shifting markets and the changing times.

Above is the company's logo and name as it stands today; at right are its logos since its founding. The earliest known logo was typical for the times — a simple scroll underscoring the letter "A." In 1962, as the company narrowed production to carbonless paper predominantly, change was in order, and a bolder, more modern logo appeared. In 1978, when the company's name changed following its sale to B.A.T. Industries, the company kept the "A" but re-imagined it into an image of carbonless paper coming off a roll.

In 1996, as the company entered the coated free sheet market, Appleton adopted a more contemporary logo intended to appeal to graphic designers and other customers for coated free sheet paper. Following the employee buyout of the company in 2001 and the post-buyout diversifications and new-product introductions, a shortened name, simply "Appleton," and logo were born. The current logo carries the tag line, "What Ideas Can Do," and the crisp "O" in the logo suggests the evolution of an idea — still the innovative heart of Appleton.

To celebrate its 100th anniversary in 2007, Appleton painted its fleet of trucks with a new logo designed for the occasion.

EARLIEST KNOWN APPLETON LOGO

1962-1978

1978-1996

APPLETON PAPERS

1996-2003

Trouble at NCR

By the mid-1970s more than 2,500 employees worked at the plants in Appleton, Roaring Spring, Combined Locks and Harrisburg and the capsule plants in Portage and Dayton. Carbonless demand was so intense that workers coined a phrase — "Get it wet and head it north!" — describing the frenzied activity. Cashman explains that "the plant ran south to north, and the sheet would start at the south end of the machine and finish up at the north end. The objective was to coat it, dry it and ship it."

Total sales of the NCR Paper brand of carbonless since its introduction exceeded one million tons by 1977. The next year, Appleton Papers' sales reached a record $330 million. The company accounted for 200,000 tons of the 322,000 tons of carbonless paper sold in the United States that year, with Mead's 70,000-ton production a distant second. Says Schumaker, "We were making money hand over fist from carbonless."

Relations with NCR were generally good, thanks to the company's impressive cash flow, which offset declining revenues in the parent's core business. Following the second Arab oil embargo in 1978, however, the U.S. economy flattened, interest rates soared into the double digits and many companies were caught in the tailspin, NCR among them. "They were in dramatic decline, which created some friction," says Paul Meier, who joined Appleton Papers' industrial relations department in 1967. "They were cutting costs, slashing the workforce and not giving wage increases, and here we were growing like gangbusters. They didn't keep up with our needs internally or the wage expectations of our employees."

Problems also surfaced when NCR entered the business forms market with its own branded product. "They made forms through a division called Systemedia that competed with other forms makers, and there was this thought in the marketplace that Systemedia was getting favored pricing from us," says Sullivan. "When a printer lost a contract to Systemedia, bad feelings would develop, and it trickled down to strain our relations with NCR."

In June 1978 the long association between NCR and the company drew to a close when NCR sold Appleton Papers for $280 million to B.A.T. Industries p.l.c. NCR had decided to concentrate on its core business, which was transitioning from mechanical cash registers to electronic point-of-sale computers; it sought an infusion of cash to invest in this strategic objective. It later was learned that NCR previously had rejected a bid from International Paper to buy Appleton.

B.A.T. Changes

B.A.T. Industries had been formed in 1976 upon the merger of British American Tobacco Company and Tobacco Securities Trust. A conglomerate with wide-ranging assets from tobacco companies to cosmetics firms, B.A.T. was the third largest industrial enterprise in England. Among its holdings was Wiggins Teape, a licensee of NCR since 1956, and the largest manufacturer of carbonless paper, under the Idem brand, in Europe. Wiggins Teape operated 10 paper mills and eight factories in Europe and produced a wider range of papers than any other company in the world. Following the acquisition, B.A.T directed Appleton to transfer its substantial European carbonless business to Wiggins Teape. Appleton Papers became a subsidiary of B.A.T. under a slightly different name than had been used in 1971: Appleton Papers Inc. (the comma after "Papers" was removed). A new logo — developed by graphic designer Ray Hudson at Adstaff, a local ad agency, and Bob Suess, Appleton's advertising manager — was unveiled. It combined the "A" from Appleton with a symbol designating carbonless paper coming off a roll.

A more substantive change took place at the top. After 43 years of service, John Reeve retired in November 1977, just prior to the acquisition and well aware of the pending sale. He was succeeded as president and CEO by John Hangen, who began his career in NCR's mailroom in 1941, rising through the finance and accounting ranks to become senior vice president of finance. Reeve's absence was keenly felt. During his career he'd been a sales representative, personnel manager, plant manager, vice president and then president and CEO. Undeterred by the risks of trying new things or making big decisions, he had embarked on an ambitious expansion program that continually kept the company technologically ahead of its peers.

Following the B.A.T. acquisition, Hangen was appointed

Sheeters like the one above at the Appleton plant cut coated sheets of paper to smaller size then slide them into a conveyor for wrapping into reams. Opposite, Mike Jensen, left, and Tim Jones Van Eyck use computer controls to measure and adjust the moisture content and weight of paper being run on the No. 13 coater.

chairman in addition to his other roles. His main goal, he said at the time, was to double carbonless sales in five years while improving profitability. So intense was the spotlight on carbonless that one decorative paper line after another was abandoned during the '80s. "When we pulled out of the shelf liner business, I got a phone call from the actresses Claudette Colbert and Joan Bennett, who said they loved the product and wanted to know where they could get it now that we no longer sold it," says Frank Sanders. "I felt so bad I sent them our last carload."

A factor in the demise of many legacy products was "the difficulty of running two different companies — carbonless and thermal and then everything else," says McIntosh. "There was competition internally for resources." When the company merged its two sales groups in 1980, it was "the beginning of the end for many commercial papers," says Jerry Thiel. "Carbonless had become so big that when a salesperson called on distributors that handled both carbonless and commercial products, since 90 percent of the quota came from carbonless, 99 percent of the time was spent selling it. Ultimately, we lost the business."

Not all coated paper products fell by the wayside. The company continued to make Currency Cover and fluorescent paper grades. Carbonless was king, followed by thermal papers.

Carbonless was king, followed by thermal papers. The company continued to make Currency Cover and fluorescent paper grades.

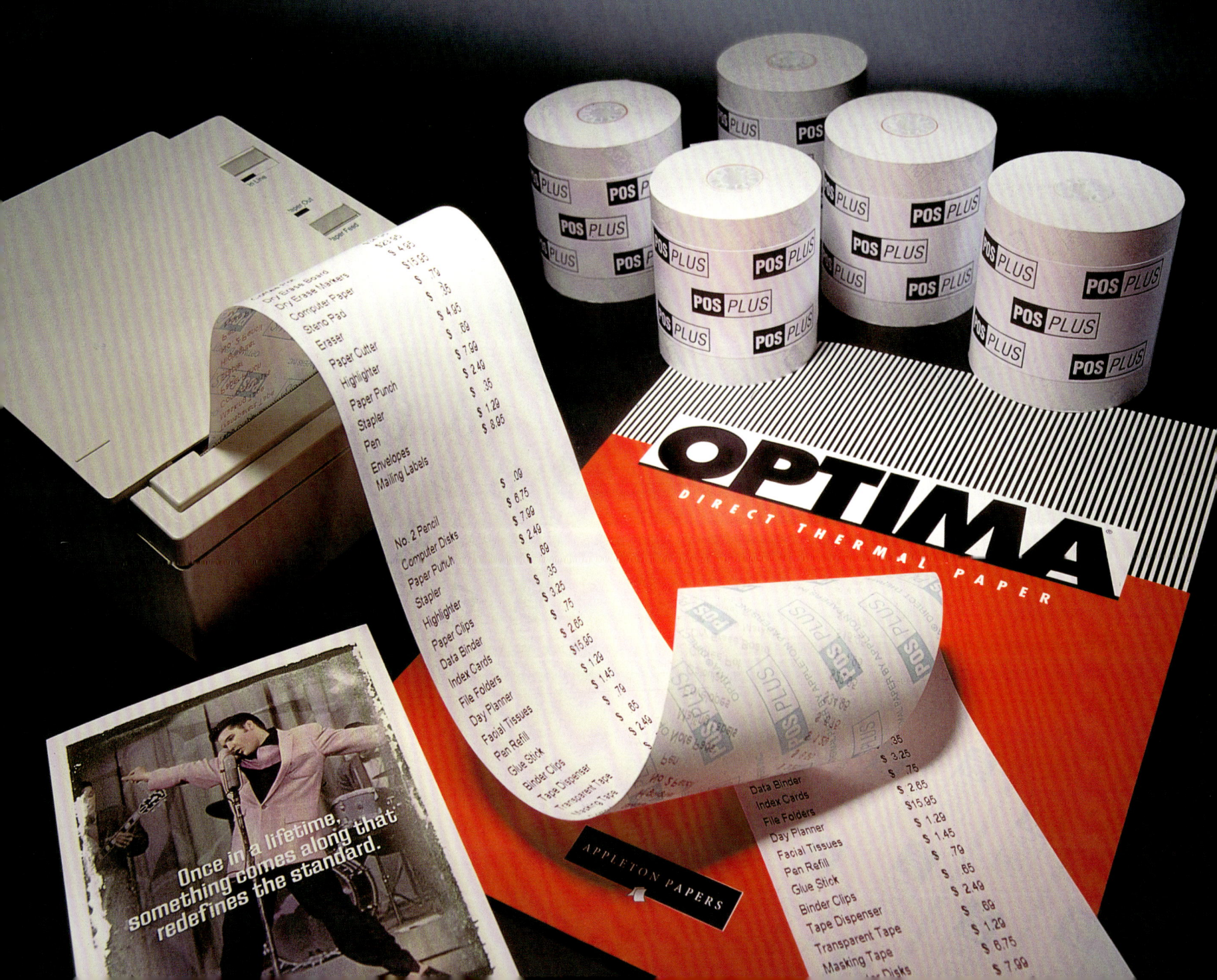

In 1980 the company also parted with the capsular products division that had furnished mood rings and other novel products, selling it to a group of investors in Europe. Microcapsules were still manufactured in Portage, but only for carbonless coatings. The Dayton capsule plant had shuttered after the company's divestiture by NCR.

The import was clear: B.A.T wanted Appleton Papers to put virtually all its efforts behind increasing its share of the North American carbonless paper market. Little wonder why — carbonless paper was a cash cow, accounting for $367.6 million in sales in 1980. "They told us to think expansively," says Meier. "They wanted us to invest in the training of our people and were willing to reward improved performance with broader compensation. Unlike NCR, they didn't pinch pennies. It was a breath of fresh air."

The parent company reorganized in 1980, creating Louisville, Kentucky-based BATUS to operate its U.S. businesses, which included Appleton Papers, Brown and Williamson Tobacco Company and several department stores, such as Saks Fifth Avenue and Gimbels. The decision put an ocean between Appleton Papers and Wiggins Teape, removing any concerns over a more formal alliance. "We operated independently and autonomously," says Schumaker. "John Hangen was enough of a corporate politician to ensure we did our thing and they did theirs. He was always thinking ahead to the next moves, and he clearly managed to get money for us from BATUS."

Appleton invested the capital principally in the Harrisburg plant, which underwent significant expansion from 1979 through 1981. The company installed two new tandem coaters plus a state-of-the-art plant to manufacture capsules at much higher productivity. "Prior to Harrisburg, we averaged one or two batches a month that were rejects, mostly because of operator error," says Don Reinhard, a former NCR manager who was superintendent of capsule operations at Harrisburg in the 1970s. "We're talking 2,000 pounds of chemicals that cost more than $10 a pound, so it was a significant loss. When we built Harrisburg we kept that in mind and introduced automated procedures that got us down to one or two rejects a year."

No longer was Harrisburg just a storage and distribution center — the addition of rewinders, material preparation equip-

Carbonless paper remains the company's top revenue generator, but Appleton Papers also continues to make legacy lines like its metallic and fluorescent grades, left. Thermal paper, opposite, an invention of several NCR scientists including Henry Baum, who later moved over to Appleton Papers, became the company's second biggest earner.

ment and packaging machinery over the years had increased the plant's size to 459,000 square feet by 1982.

Capital also was invested in improvements at the other plants, including a $37 million water treatment facility in Roaring Spring and a $30 million modernization program at Combined Locks. More sophisticated computers and automated processes made their debut, as did new products like the Rally brand of carbonless paper for high-speed copy machines. The company also unveiled differently priced categories of carbonless paper for business forms, from premium paper with enhanced color attributes called Superior to economy grades like Sprint and Stocktab. A new market emerged when credit card companies, which had long resisted using carbonless paper, finally changed their tune.

BATUS Invests

Appleton Papers' first thermal facsimile paper also took a bow in the mid-1980s. Although the fax paper suffered initially from curling, image fading and background darkening, the technical department eventually solved these problems. Thermal fax paper had a short life span, however. Fax machines using plain paper soon came on the scene and obliterated the market potential for thermal fax.

The BATUS money spigot also funded a series of acquisitions during the decade, beginning with the purchase of a 170,000-square-foot paper converting plant in Peterborough, Ontario, in 1984, from Nashua Corporation, a competitor. To

Dan McIntosh, above, served as Appleton's vice president for sales and marketing. He spearheaded the company's adoption of quality principles from the late 1980s into the mid-1990s. The company's environmental initiatives of the 1990s included the Foampak recycling program, below and opposite. Foampak was a film and foam wrapping that protected rolls of paper from damage.

market its products in the rapidly growing Canadian market, Appleton had established a subsidiary in the country four years earlier, but government regulations now required on-site manufacturing in order to sell products in Canada.

At the Peterborough plant, Appleton introduced a new wrapping for carbonless paper. Called Foampak, the foam and film wrapping protected rolls from damage, enhancing yields and performance. With the company-wide expansion of this new packaging concept, Appleton launched a Foampak recycling program to reduce waste. The company also began using recycled paper for the thermal and carbonless grades. In later years Appleton became a principal benefactor and member of the National Office Paper Recycling Project.

No sooner was the ink dry on the Peterborough contract than the company executed another major acquisition. It paid $83 million to buy the P. H. Glatfelter pulp and paper mill in West Carrollton, Ohio, south of Dayton. Glatfelter had taken ownership of the mill following its 1979 acquisition of Bergstrom Paper Company, which had acquired the mill previously from Kimberly-Clark. The facility boasted three paper machines with a combined capacity of 120,000 tons per year, although only two machines were operating at the time. "We were doing book, bond and writing grades, and had no coating capabilities," says Mark Ferguson, former West Carrollton mill manager.

Appleton converted West Carrollton's paper machines to coat both CB and CFB paper. Then, in 1987, the company restarted the No. 93 machine — the largest, widest paper machine in the world producing emulsion-grade CB at the time; that machine later began coating CFB grades, as well. Fiber for much of the paper came from de-inked waste paper.

The acquisition was the last John Hangen negotiated. In March 1984 Tom Busch succeeded him as chairman and CEO. A chemist by training, Busch had been with the company since 1948 and had worked tirelessly to solve the carbonless coating and related production problems. Busch's tenure at the top lasted little more than a year,

however. His wife Jenny's illness required that he devote more time to her needs, and he announced his retirement in December 1985.

On New Year's Day the company appointed John W. Turner chairman and CEO, and Dale Schumaker president and chief operating officer. Turner had joined Appleton in 1971 and had risen through the ranks to become senior vice president of marketing. Under Turner the company turned its attention to vastly improving customer service and product quality. Competition in the carbonless line was intensifying, and concerns mounted that a rival like Mead might steal away a top customer like Zellerbach Paper Company, the company's largest merchant account from the Rocky Mountains to the Pacific.

To get a pulse on customers' perceptions, the company's sales and marketing department surveyed merchants, printers and end users. This Carbonless Roll Attribute Study, as it was called, solicited customers to rate the company's products and service against its competitors. "What came through was that we were arrogant in terms of our pricing and flexibility," says McIntosh, vice president of sales and marketing who headed up the executive quality steering committee looking into the matter. "The days of 'get it wet and head it north' had taken a toll on customers, who objected to product defects, late shipments and scant regard for their concerns and issues."

The Push for Quality

The results of the 1986-87 survey prompted the company to undertake a strategy of continuous quality improvement that began with a series of quality improvement process seminars given by Philip Crosby Associates, Inc., a quality-improvement consultancy. Every employee was subsequently trained in the Crosby quality principles. "After the Crosby courses, we began to embrace concepts like 'zero customer complaints' and 'zero product defects,'" says

UPDATE: APPLETON PAPERS' FOAMPAK RECYCLING PROGRAM.

98 APPLETON

The West Carrollton Mill

In 1984 Appleton Papers Inc. purchased a paper mill in West Carrollton, Ohio, owned by P. H. Glatfelter Company. The location of the mill fit nicely with company objectives, as it was a short distance from Dayton, home to its former parent NCR Corporation. "We are quite familiar with the Greater Dayton area," said Tom Busch, the company's CEO and chairman at the time. Of course, the primary reason for buying the mill was the dire need to increase carbonless paper manufacturing capacity.

The mill has changed hands several times since its christening in 1948. That year, it was built by the American Envelope Company to make the obvious — envelopes — from a single paper machine. "The story goes that the company's owner Karlton Smith had trouble getting paper to make envelopes during World War II, so he built his own papermaking plant," says Keith Paden, production manager at the mill in the mid-1980s and 1990s.

In 1959 Smith sold the mill to Kimberly-Clark Corporation, which produced bond grades of paper. Kimberly-Clark installed two paper machines (No. 92 and No. 93), finishing operations and the facility's main office building. In 1972 Bergstrom Paper Company in Neenah, Wisconsin, bought the mill. Seven years later the mill fell into Glatfelter's hands when it acquired Bergstrom Paper.

The team that oversaw the rebuilding of the No. 93 paper machine at the West Carrollton plant in 1996 poses for a portrait with the machine, left. Above, an aerial view of the mill.

When Appleton Papers took ownership at an $83 million price tag, the mill encompassed three paper machines with the capacity to produce 110,000 tons per year. Primary grades produced included tablet, bond and offset paper. Converting the machines to make carbonless grades was a top and immediate priority. Machine No. 91 was converted in 1985, followed by No. 93 in 1987 and No. 92 in 1991. Paden calls the ordeal in converting the first machine a "train wreck."

He adds, "Glatfelter had put no money into the plant, and it was poorly maintained. It took nine months just to get No. 93 to work, much longer to convert it. The whole place was a disaster. We started at the ceiling, painting it, and worked our way down to the floor, sandblasting everything along the way. Nos. 91 and 92 also were in bad shape, but not nearly as bad as No. 93. At least they were operable."

Tom Sullivan, a retired sales administration manager, says "Glatfelter had essentially shut the mill down before we bought it. Our agreement was to get the machines running with a few orders to provide employment to a skeleton crew of mill workers. I was sent there to keep the machinery running. It was a scramble, and we struggled. I remember we made printout paper for computers for a year or more just to keep the employees working and to allow time for the conversion to carbonless."

In March 1999 the West Carrollton mill became one of the country's first paper mills to receive ISO 14001 registration. The registration signifies that the mill meets a series of environmental management standards for companies around the world because the mill operates with a verifiable system to minimize effects on the environment caused by its activities. The Roaring Spring mill and Appleton Plant have since also received ISO 14001 registration.

Today the mill, managed by Satish Damadaran, employs 410 people. It has three paper machines and recycling operations that process almost 400 tons of waste paper every day — some of it broke paper from the Appleton and Roaring Spring facilities and the remainder coming from the waste paper market.

In January 2007 Appleton announced a $100 million expansion of the West Carrollton mill to add a state-of-the-art coater to produce thermal paper and to construct related facilities and enhance the mill's No. 92 paper machine. CEO Mark Richards said the expansion, which is scheduled for completion in mid-2008 and will add 35 jobs to the mill's workforce, represents "a commitment to reinvest in our core business when we have attractive opportunities to do so."

Tom Cashman, who attended the seminars. "Although we might not reach these goals, the idea was to always chase them."

Previously, it was not uncommon for one department to turn a blind eye to a product problem and simply pass it onto the next department. "Tonnage was the only metric that counted, and workers did all they could to meet tonnage expectations," McIntosh explains. "We decided to make a change and recognize accomplishments not on overall tonnage, but on tonnage of perfect product that met a specific on-time delivery schedule. People now began thinking about what they could do to make the next guy's job easier."

The company's new corporate theme said it all: "Think Quality — Do it Right the First Time."

Appleton adopted other modern production concepts, including MRP, for Manufacturing Resource Planning. "The silos separating everyone were torn down, and we began to implement computerized systems integrating everything from order entry through sourcing through production and shipping, always with an eye to the customer," says Casey. Personal computers soon became a desktop fixture, leveraged to study market trends, solve business problems and improve overall decision making. In May 1986 the company christened a 16,000-square-foot Computer Data Center on North Rankin Street housing 55 information services employees. The effort had been led by John Tucker, the company's first vice president of information services.

Ironically, the presence of computers sounded alarms in the carbonless paper market by shifting business transactions to automated paperless systems and, in later years, to Internet-based commerce. After years of sky-high production rates, Appleton Papers realized that the carbonless paper market would gradually decline. To compensate for a potential deterioration in revenue and profit, the company would have to continually increase its market share. As Turner put it, "If we can make a true shift ahead of the market in quality, we can gain a competitive advantage."

This strategy became even more important following a series of marketplace events that jeopardized the company's dominance in carbonless paper. In 1986 Mead paid $250 million to acquire Zellerbach, the paper merchant that had constituted nearly half a billion dollars in annual sales for Appleton Papers. International Paper also was in the midst of creating its own merchant distribution arm through acquisitions. And Appleton's number-one distributor, Boise Cascade, announced that it would now market a carbonless product of its own. The company had no good strategic alternative but to terminate its relationships with both Zellerbach and Boise Cascade. "We needed alternate distribution, and fast," says McIntosh.

Acquisition and Expansion

Over the years, Appleton Papers had developed a close relationship with a growing merchant operation in Philadelphia, Alco Standard, headed by Ray Mundt. "Ray had this grand vision of putting together a coast-to-coast chain of smaller merchants around the country, and, under the leadership of Pike Peterson, he got the ball rolling in the West, acquiring a few companies and calling the combined organization Unisource," says McIntosh. "We went to Pike and offered him the opportunity to launch Unisource with the NCR Paper brand. He didn't think we were serious. It turned out to be the biggest merchant move ever to happen in the industry. Together we converted all of the Zellerbach business to Unisource and actually gained market share." Unisource today is among the largest paper merchants in North America and one of Appleton's largest customers.

The acquisition spree that had begun under Hangen continued under Turner. In 1986 Appleton Papers purchased Muskegon, Michigan-based East Shore Chemical Company, a manufacturer of industrial organic chemicals. Appleton heretofore had bought the specialty dye used to make carbonless paper from two companies, Hilton Davis in Cincinnati and Yamamoto in Japan, but felt increasingly cornered by capacity and price demands. "We were paying about $16 a pound for the dye, and our research chemists were convinced we could make it ourselves for a lot less," Schumaker explains. "They were right, and we were soon manufacturing dye at about 10 bucks a pound." The acquisition was a rare vertical integration move by the company.

Other company expansion activities also were in progress.

During the 1980s Appleton poured capital into upgrading and rebuilding plants and machinery. At the Appleton plant, the No. 13 coater, left, was installed and went online in 1988 to produce thermal paper, which had developed into a growing market. The same year, the plant also boasted a new automatic storage and retrieval system accommodating more than 1,000 rolls of paper.

Gordon Bond, above, CEO from 1989 to 1993, led the company's push to adopt quality as a core management discipline. The company has twice been a finalist in pursuing the prestigious Malcolm Baldrige National Quality Award — in 1994 and 1996.

At the Appleton plant, the No. 13 coater was installed to produce thermal paper, a $16 million "space utilization project" was undertaken to improve material flow, and an automatic storage and retrieval system to accommodate nearly 1,200 rolls of paper was implemented. Pulp manufacture was improved at the Roaring Spring mill, and two machines at the West Carrollton mill were overhauled to coat CF paper. At Combined Locks, the No. 1 paper machine was rebuilt to deliver a 60 percent increase in production.

Since acquiring Appleton Papers, B.A.T. had invested some $600 million in fixed assets, the acquisition of additional paper mills and the rebuilding of all the production lines. The generous parent had assisted enormous growth, and the company was now three-and-one-half times the size it was under NCR.

Gordon Bond became Appleton's chairman and CEO in April 1989, following Turner's retirement. Bond had been CEO of Wiggins Teape's carbonless operations in England, and was considered a sales and marketing guru. "Gordon refocused the company toward sales and marketing and put a renewed vigor behind our quality initiatives," says Schumaker. "Under Gordon we increased market share significantly."

Well aware that the carbonless market was maturing, Bond was nevertheless convinced that the company could achieve additional growth, profit and renewal. In his first management conference, entitled "How to be successful in a mature business," he provided the solution — differentiation in the eyes of customers. As he recalls today, his determination was grounded in one belief: "Great things were still possible at Appleton."

Spinoff Fends Off a Raider

The new CEO created separate strategic business units for thermal paper and coated paper to focus efforts on the development, production and sales of each line. He also engaged several consultancies to help the company capitalize on its opportunities. Working with The Mac Group, each plant focused on specific products and processes to assist quality and service imperatives. Bond also expanded the distribution system and invested generously in information technology systems and software.

The strategy hit a bump in 1989 when Sir James Goldsmith, a notorious corporate raider, made an unsolicited tender offer to buy B.A.T. Industries. Goldsmith's goal was to dismantle the conglomerate into its corresponding components and then sell each of them off for large gains. Although B.A.T rejected the hostile takeover attempt, it was forced to reorganize to satisfy shareholders. B.A.T. put Appleton Papers and Wiggins Teape into a new holding company and then spun the new entity out as an independent company called Wiggins Teape Appleton.

In December 1990 Wiggins Teape Appleton merged with the French paper group Arjomari-Prioux. The new company was named Arjo Wiggins Appleton, or AWA. Each of AWA's three constituent companies operated independently, though Appleton Papers was the main profit contributor.

These startling developments compelled Appleton Papers to rationalize its holdings to focus on core competencies and key markets. It shut down the Peterborough facility in 1990 and moved that plant's carbonless coater and associated equipment to Appleton. The next year, it sold East Shore Chemical Company, the company's dye-producing plant, for $40 million to Yamamoto Chemicals.

Appleton redirected the proceeds into major plant improvements and hoped-for acquisitions, including failed efforts in 1990 to purchase a paper mill owned by Boise Cascade in Vancouver, Washington. "We needed paper and pulp capacity," says Schumaker. "As part of the deal, Boise promised to provide long-term pulp shipments to Vancouver from its other facilities. We cut a deal, but then the government got into the middle of it. In April 1991 we walked away from it, and Boise shut the plant down." Although a disappointment, the failed deal had a silver lining: Appleton won most of the business Boise abandoned.

Instead, Appleton earmarked more than $170 million for a massive expansion of the Combined Locks mill, including the installation of a new paper machine, a new gas-fired boiler and additional support facilities and infrastructure. The successful project ultimately increased the mill's annual production by more than 100,000 tons. Simultaneously, the company undertook a $35 million expansion of the Appleton plant that included the installation of a new thermal coater (No. 17) and related support facilities. In 1996 the company invested $26 million to rebuild the No. 93 paper machine at West Carrollton to enhance carbonless paper production. The company also announced upgrades to some paper machines at Roaring Spring and Combined Locks.

Customer Focused Quality

By 1993 the company had grown to a size never imagined in the days of Charlie Boyd and Fred Heinritz. It boasted six manufacturing facilities across the country, 30 sales offices throughout North America, 10 distribution centers and a workforce of 3,600 people. Appleton Papers was the U.S. market leader in thermal paper production for facsimile, tag, ticket and label applications, and the global leader in carbonless paper production, with market share increasing in both.

While capital from its parent companies through the years had assisted this escalation, a chief factor in Appleton's growth was the impact of the company's emphasis on customer service and quality, an initiative that continued under McIntosh. "I contacted a company in Milwaukee to address our annual sales meeting because they had previously introduced me to the Customer Focused Quality concept," McIntosh recalls. "After the CFQ presentation, Gordon [Bond] became convinced this was the way for us to go."

CFQ calls for every process within an organization, from manufacturing to delivery, to be designed to ensure the most positive customer experience — and it redefines the notion of a customer to include anyone, inside or outside the company, who is the recipient of an employee's work. "Employees are required to know their customers and their expectations, whether or not those expectations were met and the ways in which we can get better at satisfying these expectations," McIntosh explains. Toward this end the company created a process called BIFI — for Bright Ideas for Improvement. "We encouraged employees to contribute their thoughts on continuous improvement," McIntosh adds. "In the first year we received more than 7,000 ideas."

Paul Meier, who was vice president of human resources at the time, says employees "became energized around the concept of serving customers. We even had license plates with the letters `CFQ' on them. … CFQ became our rallying cry." In 1993 Appleton hosted the first Conference for Quality Achievement at the Roaring Spring mill.

Appleton incorporated many modern manufacturing methodologies to facilitate the change from a production-driven company to a customer-focused business. These included Just-In-Time principles to manufacture and deliver products on demand. In all respects, Appleton became a highly sophisticated and very modern organization.

These various initiatives coalesced behind a major corporate objective — winning the prestigious Malcolm Baldrige National Quality Award given by the U.S. Department of Commerce. The award was at the time considered the ultimate recognition of quality-focused management and customer service. McIntosh led the Baldrige effort, mobilizing the entire organization to prove that Appleton was the industry leader in quality — "to walk the talk," as he says.

In 1994 Appleton earned a prestigious site visit from the Baldrige Award examiners. Although the company didn't win the competition, it was one of six finalists in a field of 71 applicants. "We gained invaluable feedback by participating in the application and site-visit process, measuring ourselves against the most rigorous standards of quality achievement in the world," McIntosh says. "It served as a road map for our continuous quality improvement." The company pursued the Baldrige Award again in 1996; once again it was a finalist.

During his 31-year career with Appleton, Dale Schumaker served in several manufacturing management positions. He was named company president in 1986 and became CEO in 1993. As CEO he presided over the company's first Conference for Quality Achievement in 1993.

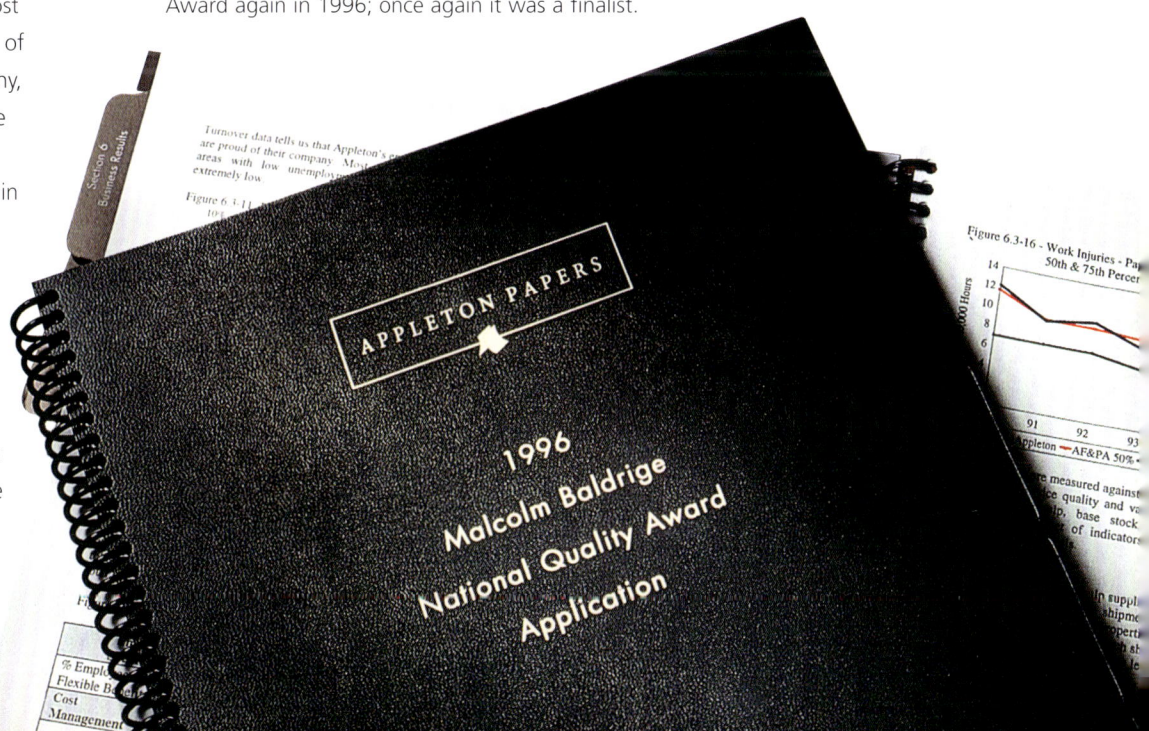

Customer Focused Quality

In the early years of producing carbonless paper, demand was so pronounced that plant workers chorused, "Get it wet and head it north!" a reference to coating the paper quickly to get it out the door. By the mid-1980s this speed-at-all-costs approach had taken a toll on customer perceptions, driving the company to continually improve product quality to effectively serve customer needs.

This effort began in 1986 under CEO John Turner, who urged plant workers, "If you know what can be done to improve quality, do it now! I really mean it. … If you know that quality isn't there, shut the machine down. Don't produce any more." Turner realized that product quality and customer service were essential tools in the effort to continually increase market share. "Service is not what we think it should be," he said. "Service is what the customer thinks it should be."

Turner laid out three strategic goals: quality, service and low-cost production. The company's target was 100 percent conformance to manufacturing specifications dedicated to serving the needs of customers, both internal and external. Strict observance to on-time delivery of products and services was a principal objective, as was adherence to the maxim, "Think Quality: Do it Right the First Time."

To sharpen the focus on these strategic objectives, Appleton Papers invested more than $600,000

Appleton introduced an unconditional guarantee in conjunction with its commitment to quality improvement. Left, employees at the company's 1997 Conference for Quality Achievement.

to send its executives and managers to a series of Quality Improvement Process training seminars given by Philip Crosby Associates, Inc. Executives learned the benefits of creating quality-improvement teams and the need to continually remind and educate employees about quality and service initiatives.

The Crosby training seminars were just the first step. "We learned a lot of good principles about quality, including process, measurement and systematization, but the training ultimately was too programmatic and formulaic and didn't have enough feel for the customer — the soft side of the business," says Dan McIntosh, who took the courses and later headed the company's steering committee on quality. "Nevertheless, it was a solid foundation upon which to build more robust quality initiatives."

The next stage in the company's quest for quality occurred in the early 1990s under CEO Gordon Bond. Bond started Appleton down the path to later apply for the prestigious Malcolm Baldrige National Quality Award, awarded by the U.S. Department of Commerce to companies possessing superior product quality and customer service. "I was at Xerox for 10 years, and we went for and achieved the Baldrige," says Bond. "I believed this was a prime way for Appleton to differentiate itself from our competitors in a mature market." Twice Appleton aspired to win the Baldrige, and twice it was a finalist, an important achievement.

Bond gave every employee a gold card explaining the quality strategy, which he called "Customer Focused Quality," or CFQ. "We were bringing hundreds of people from all across the organization together to talk quality, really discussing what it is and how we could continually achieve improvements," says McIntosh. "Gordon saw what we were doing and was never one to shy away from a good idea. What he brought to the effort was his willingness to take charge of it, not to mention his inventiveness." Bond's marketing cleverness even spawned employee license plates adorned with the CFQ logo.

CFQ is the company's overriding philosophy, the single most important strategy undertaken in the modern era. Initially it was its own separately managed initiative. In the 1980s and 1990s CFQ became integrated into the way Appleton did business. Today it is being redirected into a market-facing program to help the company be better than anyone at understanding and addressing customer needs. Through CFQ the company has grown market share, achieved world-class levels of customer satisfaction and built high levels of trust and confidence with customers. The philosophy is defined simply as "meeting or exceeding customer expectations in all aspects of every transaction."

Videos of various CFQ conferences over the years, including the first CFQ Conference for Quality Achievement at Roaring Spring in 1993, reveal a high-octane event, one that several employees call "evangelical." "We'd get extremely pumped up at the quality conferences," says Paul Meier, retired vice president of human resources. "The idea is to chase performance excellence continually, with the notion that this journey will never end."

It was back to the future for Appleton Papers in the mid-1990s, when the company invested in the coated free sheet business, the industry that had defined Appleton's early market success. At the Combined Locks mill, the company invested more than $170 million to install the No. 7 paper machine, a state-of-the-art machine that served as the cornerstone of the company's entry into the coated free sheet market. It is shown, opposite, in a photograph from February 1994.

In and Out of Coated Free Sheet

New Chairman and CEO Dale Schumaker, who succeeded Bond in February 1993, capitalized on the changes wrought by CFQ and the Baldrige Award competition to guide the company into another market. Ironically, the planned entry into the coated free sheet paper market to manufacture high-quality paper for book publishing, annual reports and advertising brochures marked a return to the company's legacy specialty papers business — a market it had largely abandoned. "We were going back to our roots, although the technology had changed dramatically in the intervening years," Schumaker says.

The first foray into this market was an unsuccessful attempt to acquire S. D. Warren, a major manufacturer of coated printing papers. "We came in second place on the bid," Schumaker sighs. "Consequently we had to turn our attention elsewhere, spending $170 million to install the No. 7 paper machine at Combined Locks to make base paper for a variety of coated papers. We also bought a small mill in Newton Falls, New York, [from Stora AB for $60 million] that made coated papers."

The acquisition was a letdown, a consequence of too little, too late. "We really needed to acquire S. D. Warren to enter the market with a bang, but our parent company was less than interested in putting up the kind of money to win the bid," Schumaker explains. "Without it, we faced an uphill climb."

Other executives of the time agree. "AWA felt that S. D. Warren was too big a bite," says Richard Curwen, a former finance executive from Wiggins Teape who would succeed Schumaker as CEO in 1995. "They were uncomfortable entering a market in the U.S. that was almost equal in size to our carbonless and thermal business."

Appleton Papers spent nearly $400 million at Combined Locks and Newton Falls through 1998 to become a player in the coated paper market. "We had great difficulty turning the Newton Falls mill around," Curwen recalls. "We had overrated its capabilities, and it turned out to be much too small an entrée into the business."

A scant few years after AWA became the parent of Appleton Papers, rumors circulated that AWA wanted out. "They wanted to sell us to a large paper company for whatever they could get, but few were interested," Curwen says. "A road show commenced to drum up interest among venture capital firms, and several lined up to buy the company. But, following a dip in the value of AWA's stock and increasing turbulence in the stock market as a whole, the luster came off the rose, and the potential suitors looked elsewhere."

Things only got worse. In the first half of 1998 Appleton Papers' sales fell 2 percent and its profits 19 percent. Unable to support AWA's objective to sell Appleton Papers, Curwen regretfully stepped down as president and CEO in November 1998. He was succeeded by Douglas Buth, who had worked in finance at BATUS and who had scaled the corporate ladder at Appleton Papers through the planning, sales and marketing departments. At the time of his appointment, Buth headed the carbonless business as executive vice president.

His first months as CEO were a baptism by fire. Beset by declining carbonless paper demand, Buth was forced in 1999 to close the Harrisburg plant, "a brutal but necessary move," says Cashman. Two of the plant's free-standing coaters were relocated to Appleton and most of the finishing equipment was moved to Roaring Spring. The plant was sold in 2001 and turned into a distribution facility.

The pressures mounted in January 2000 when AWA transferred the assets and operations of Appleton Papers' coated free sheet business to a new legal entity, Appleton Coated LLC in Kimberly, Wisconsin. AWA then managed this company as a single business with its coated free sheet operations in Europe and North America. Consequently the Combined Locks and Newton Falls mills no longer were part of Appleton Papers. Finally, after trying and failing to sell the Newton Falls mill, AWA made the difficult decision to close it. Appleton was effectively out of the coated free sheet market.

In July 2000 Worms & Cie, a Paris-based investment group that owned a controlling interest in AWA, purchased the balance of AWA. Worms & Cie was resolved to divest Appleton Papers. "They planned to sell us to a venture capital firm that would try to get as much cash as possible out of us, without putting another dime into the company," says Buth, his hackles rising. "Meanwhile, no paper company was interested in buying us. There were few options, and our future looked bleak."

Then the dark clouds parted.

Second Renaissance

Chapter 5

By 2001 Appleton had spent 30 years under another company's control. That stood to change when management proposed an employee buyout. Would workers believe in the company's future strongly enough to invest a majority of their retirement funds in a purchase? Below, the offering packet Appleton sent to employees prior to the vote.

Doug Buth's leadership skills were evident early on at Appleton Papers. Although he'd originally applied for a job in finance — he's a CPA with an accounting degree from Notre Dame, on whose football team he played — he found his calling in corporate strategy, sales and marketing, rising through the ranks to head the carbonless business.

Like many athletes, Buth is poised, confident and competitive. The 6-foot, 4-inch native of Green Bay, Wisconsin, played tight end for the Fighting Irish and famously caught a pass for a two-point conversion in the final minutes of a game from a quarterback named Joe Montana. Montana later led the San Francisco 49ers to four Super Bowl victories, earning the nickname the "Comeback Kid."

Buth now had the handoff from the parent company to develop a comeback strategy at Appleton Papers. He outlined his key objectives — continued emphasis on Customer Focused Quality, rapid new-product development, extending share in existing markets and implementing a sophisticated information-technology platform to drive business performance and productivity. With regard to the latter, he announced the first phase of a three-stage $23 million enterprise resource planning platform, dubbed Project Venture, in 2003.

Buth's strategic plan — GO!, which stood for Growth Opportunities — made its debut against a backdrop of enormous uncertainty. As executive vice president, he had been a member of the unsuccessful road shows hawking the company during the tumultuous stock market period of the late 1990s. Now CEO, he was cognizant that AWA's eagerness to sell Appleton Papers had not diminished. Equally nettlesome was the loss of the company's coated free sheet assets and the contraction in the carbonless market. "It was not a great situation in terms of our future," he remembers.

There was one option that held some promise, but the chance of its success seemed remote — an employee buyout of Appleton Papers. Buth was made privy to the idea by Vice President of Finance Ann Whalen, although the concept dated to 1994, when it was proposed in a memo by Treasurer Dick Wehrel to former Chief Financial Officer Rick Curtis. Buth resurrected the notion in 2000 and approached AWA Chairman Ken Minton with the idea. He was turned down flat. The following year, when Minton was succeeded as chairman by Luca Paveri-Fontana, Buth took another stab. "I told Luca that we had had a divestiture process that had gone nowhere and that this was not likely to change," he says. "I then proposed an ESOP (employee stock ownership plan) as an alternative." The message did not fall on deaf ears.

Putting a Deal Together

Buth enlisted the assistance of Houlihan, Lokey, Howard & Zukin, a Los Angeles-based international investment banking firm, to further develop the plan. Lou Paone, the firm's managing director, laid out the structure, benefits and risks of an employee buyout to Paveri and other AWA executives in London. "Lou explained a novel structure in which funds transferred from the employees' 401(k) retirement plans would fund a portion of the sale, with the rest coming from bank debt and bonds," Buth notes.

On February 12, 2001, AWA signed a letter of intent to sell Appleton Papers to its 2,500 employees. Two days later Buth announced the plan to employees. "There was mixed reaction, with some employees skeptical, some excited and others sitting on the fence not sure what to make of it," recalls Kerry Arent,

Appleton's human resources director.

By July 5 Buth had a definitive agreement with AWA to sell the company for $810 million. What lay ahead was an extraordinary amount of work, agonizing tension and high drama — not that Buth was deterred. "It was the coolest year of my life," he says.

The structure of the employee buyout was indeed groundbreaking. Tax laws regarding ESOPs had recently changed to allow an ESOP to operate as an S corporation, an entity granted special tax status under the U.S. tax code. An S corporation is generally exempt from federal income tax, except on certain capital gains and passive income. By becoming an S corporation, the ESOP would avoid double taxation — once to the shareholders and a second time to the corporation.

It was a marriage made in heaven. "An ESOP is a tax-exempt retirement plan and S corporation is a tax pass-through entity," Paone explains. "Tax is not paid at the corporate level but passes through to the shareholders, which in this case would be a tax-exempt entity. Consequently the company would pay no taxes on earnings, saving it millions of dollars and improving cash flow."

Deloitte LLP, a consultancy specializing in tax law, advised on the tax issues surrounding an ESOP operating as an S corporation. "The tax benefit of not having to pay any current income tax was extremely beneficial in terms of the transaction attracting financing," says Helen Morrison, a Deloitte principal in Chicago. "It was also extremely beneficial for employees in that there would be no crunch on cash flow, since the debt would be paid off gradually by the company in the future."

If the buyout were successful, Appleton Papers would become one of only a small number of companies where an ESOP was the vehicle for a leveraged buyout using employee 401(k) funds. "The trick now was to go to the capital markets and convince investors to co-invest alongside the employees," says Paone. "Of course, employees would need to buy in, as well. Whether any of them would invest their money was open to question, given the structure's uniqueness."

To solicit financing above the employee-buyout dollars, an amount gauged minimally at $100 million, Buth engaged investment bankers Bear Stearns. "We committed to provide the financing to get Arjo Wiggins to sign up to sell the company," says Mark Bernstein, Bear Stearns' senior managing director. "That meant we would be on the hook if we couldn't raise the money in the capital markets. Certainly this was a risk since our investment base wasn't familiar with the unusual structure."

There were other challenges, including the position of the U.S. Securities and Exchange Commission. In a typical leveraged ESOP structure, a company invests all the money required, its employees put up none of their retirement assets, and the stock in the ESOP is gradually provided to the employees over time. In the company's planned employee buyout, Appleton Papers would put up no money, the employees would use their retirement assets to buy the company, and the stock would be provided to them once the transaction closed. "We were unsure if the structure would fly with the SEC," says Chris Noyes, a

Appleton's ESOP transaction team for the employee buyout comprised, left to right, Dale Parker, Paul Karch, Kerry Arent, Doug Buth, Steve Kula and Dick Wehrel.

partner in the law firm Godfrey & Kahn in Milwaukee, which provided legal counsel. "Knowing this had not been done before, we wrote a letter to the SEC asking for guidance, and received a seminal letter back approving the strategy."

Other question marks remained. Among them was the degree to which Appleton could go in communicating the benefits of the buyout to employees, given that the ESOP would be funded by their 401(k) retirement savings and that companies have strict obligations under the Employee Retirement Income Security Act to protect retirement assets. "Most ESOP plans are compensatory, in which the company is a contributor and nothing is required of employees," says Pete Prodoehl, vice president of Benefit Consultants Incorporated (BCI), a third-party ESOP administrator that has since been purchased by Principal Financial Group. "In this case, we had to put together a plan that would invite employees to take their 401(k) dollars out of a diversified portfolio and put roughly 75 percent of it into a single stock, which defies conventional financial wisdom. And we had to do it in a way that it would not be perceived as threatening their jobs or otherwise coercing them."

Consequently management needed to observe strict boundaries in its discussions about the buyout with workers, a departure for a company that had prided itself on open and honest communication. To assist this objective, the first of many weekly newsletters called *Ownership Updates* came off the presses in March 2001 to apprise employees of official information regarding the ESOP strategy.

Few believed that employees would invest 75 percent of their retirement savings to acquire Appleton Papers. "It was a huge amount to expect employees to contribute," says Steve Kula, the company's controller. "But management knew it was the right thing to do and didn't shy away. All of us were so convinced this was the way to go we agreed to commit 100 percent of our 401(k) savings to the buyout. Appleton Papers needed to take control of its own destiny."

Heading up the acquisition project internally was Paul Karch, a Harvard-educated attorney who was vice president of the company's law department. "Once we had the letter of intent from AWA, we hit the road to talk with the capital markets, and I did much of the explaining about the uniqueness of the deal," Karch recalls. "We also had to coordinate all the different parties and figure out how the pieces would fit together, as well as clean up some legacy problems that potentially stood in our way."

A particular fly in the ointment was possible environmental liabilities emanating from NCR's prior use of PCBs in making carbonless coating; PCBs were alleged to have contaminated the Fox River. Appleton management now negotiated a contract with AWA whereby the Franco-British parent would cap Appleton Papers' liability at a manageable $25 million.

Once the banks lined up to provide the remaining $800 million to finance the deal, senior managers put long hours of effort into developing the prospectus, the disclosure document detailing the ESOP's structure, benefits and risks for employees. They were on a fast track due to AWA's desire to close the deal as soon as possible. The ESOP offering to employees ran a single month — July 25, 2001, to August 24, 2001, with the closing date set for September 30. "It was grueling," Kula remembers. "We had to provide in the prospectus all our past financial statements, risk factors and forward-looking information. We had to explain what the ESOP was and how employees could put money in it and pull it out. We could leave no stone unturned."

"The process was incredibly exhaustive," Karch says. "We'd be doing things like moving a comma two words to the right to ensure the right meaning came across. We all sat around big tables — the classic stereotype — in Milwaukee and Chicago working 18 hours a day. We had our legal and financial advisors there to make sure everything we disclosed was appropriate and in compliance. It was some of the most detailed work I've ever done."

When the disclosure document was finished, it was distributed to employees. "Once we had it in their hands, they essentially had all the information they needed to make an informed decision whether or not to invest in the company," Prodoehl explains.

Representing the employees' interest in the transaction was State Street Global Advisors. "It was our job to negotiate with the sellers to ensure employees got a fair shake," says Kelly Driscoll, the Boston-based firm's senior managing director. "We were appointed the fiduciary to make sure the transaction was fair from a financial perspective to the ESOP. And it was. Employees would not be paying more than the fair market value to buy Appleton Papers."

Nail-Biter Election

Why might employees want to buy the company? First of all, there was the employees' strong sense of pride in the work they had been doing along with their loyalty to the company and their community. "The average length of employee service here was more than 16 years," says Buth. "There were strong allegiances to this company." Then there was the company's cash flow. Even though the carbonless market was contracting, the company generated a tremendous amount of cash, capital that could be directed into other profitable businesses. Finally, there was Buth and Appleton's management team — highly regarded and effective leaders whom employees knew, respected and trusted.

The presentation team, which included Buth, Karch, Driscoll, Paone, Prodoehl, Arent and Rick Braun, a principal at Willamette Management & Associates, was highly visible during the two rollout visits to each major facility. Since the investment opportunity involved retirement money, employees were invited to bring their spouses and financial advisers to the meetings. At every stop the transaction team was greeted by large, well-informed audiences.

"The first meeting we had with employees was in Roaring Spring," Driscoll recalls. "I remember being struck by the sophistication of the questions asked; it was clear that many workers had read the prospectus closely. People had queries about the revenue projections and the debt service, since the company would be taking on a lot of debt. Other questions concerned the structure of the board and executive compensation. Employees were extremely prepared."

At the rollout visit to the Appleton plant, Sally Feistel, vice president of the local union at the time, says "everyone was asking everyone else what they would do — would they vote `yes' or `no.' This was a very tough decision for all of us."

To help employees in their deliberations, BCI set up a hotline staffed by operators who took calls each day until 10 p.m. When there was a high volume of calls, "we'd all answer the phones here," says Cindy Prodoehl, a communications consultant with BCI and the wife of BCI's Pete Prodoehl. "Everywhere Pete and I went in town, employees would ask us about the deal and if we thought the buyout would work. Every day of the ESOP offering was intense. Pete and I became very emotionally involved."

The *Milwaukee Journal Sentinel*, below, heralded the news of the employee buyout.

"The success of any business comes down to the people who own it and the people who work for it." Fortunately, at our company those groups are the same. ... Thank you for investing in our company. We are the best possible owners." — Doug Buth

As CEO, Doug Buth, above, championed the employee buyout of Appleton Papers. He joined the company in 1988, was named CEO in 1998 and retired in 2005.

According to an essay later written by Stephanie Essex Elkins for the Wisconsin Department of Financial Institutions, "The actual election period was a nail-biter. There was no way of predicting the outcome, and one of the significant concerns was the economic climate of the time. Leading up to the election, the [stock] market was slowing. The company knew they needed $100 million from the employee retirement funds to make the deal work, and when they started, the funds were worth about $160 million." By the time of the vote, they were worth $135 million.

The ballots trickled in during the election period, and BCI tallied them. Employees had the option to designate anywhere from zero to 100 percent of their retirement funds to the buyout. "Two days before the election closed, we had about $60 million, and I thought we'd never make it," says Pete Prodoehl. "I got a call on the hotline from an employee who said he'd never received his ballot. We tracked it via FedEx, and it was on the fellow's porch. `So that's what I've been stepping over,'" the guy said. I urged him to get his ballot in immediately."

By the 28th day, BCI counted $80 million in investments — $20 million short of the total needed. While reaching the goal remained a daunting task, Buth's confidence in the buyout strategy and its appeal to company employees never wavered. "He truly believed the buyout was in everyone's best interests," Pete Prodoehl says.

Halfway through the final day, the Prodoehls got out their calculators. Cindy called Buth on the phone. "She was crying," he recalls. "I said, 'We made it, didn't we?' and she said, 'Yeah, we did.'"

Roughly 90 percent of employees voted to contribute an average of nearly 75 percent of their retirement savings to buy stock in the company. The total investment added up to $106.8 million, 73 percent of the value of the 401(k) funds. Sally Feistel gave 100 percent of her retirement money to the transaction. "My financial advisor was against it on the grounds that I'd be putting all my eggs in one basket," she says. "It was a gamble, I knew, but I went against him. Obviously many other union members felt the same way. We trusted in the future of this company and in its management."

A week and one-half later, the employees' retirement savings were converted to cash and parked in a secure money market fund. "The stock market was falling because of the technology bubble, and we wanted to preserve as much value as possible," Karch explains. "At the same time, we were negotiating with AWA's team of lawyers and bankers on the many details of this very complex transaction. Dick Wehrel, our treasurer and a tax genius, negotiated one of the best terms of the deal. He worked the taxes with the parent company and was able to get us an extra $25 million, which was great planning."

The time had come to visit Bear Stearns in New York to secure the remaining funds to close the deal. While the bank loans were in place, the mezzanine layer of bond financing was not. "We were rolling out the bond financing and had planned a meeting with Appleton management on September 14," says Bear Stearns' Bernstein. "Three days prior, of course, everything stopped." Terrorists had attacked New York.

Several executives involved in the deal were in Manhattan the day two hijacked planes flew into New York's tallest buildings. "Chris Noyes and I were preparing to meet with Bear Stearns, and on this little TV in the elevator it said the World

Trade Center had been hit by a small plane," Karch recalls. "I called Doug in Appleton from my room, and he told me it wasn't a small plane. And then the second jet hit."

"The magnitude of the event came in waves," says Noyes. "We were in the midst of this terrible human tragedy and weren't thinking at all about the financial markets. The next day, a Wednesday, the stock exchange was closed, and we tried to get a meeting going, but it didn't happen. On Thursday we were sitting in a conference room with our counterparts at the law firm and about 20 others involved in the deal when we were told a credible bomb threat had been called in. We piled into the elevator and exited the building. I distinctly remember Paul whispering, `Are these the last people we'll ever see?'"

Cognizant that not much business could be done in the aftermath of the disaster, the Appleton contingent, which included Wehrel and Noyes' partner Jim Phillips, was planning to drive home when they were apprised that a jet had been chartered for them in Glen Falls, New York. They hopped on an Amtrak train north and eventually made their way back to Wisconsin.

The stock market plummeted in the wake of the terrorist attacks, with the Dow Jones Industrial Average falling 17 percent from September 10 to September 24. "Had we not converted the $106.8 million in employee retirement money to cash just a few days earlier, it would have been worth $89 million," Karch notes. "It's questionable whether or not we would have been able to do the deal."

Anxiety reigned in the weeks following the catastrophe. "We kept asking ourselves, `What about the bonds?'" Karch says. "The bank debt was okay since Bear Stearns believed in the deal, but it was a bit fuzzy if we'd be able to do the bonds." Once the airlines reopened for business, Buth; Dale Parker, Appleton's chief financial officer; and others revived the road show, but it was abundantly clear they would not be able to float the debt to close the transaction. "We were the only ones calling on the bond market at the time; Doug and Dale would be sitting alone in the lobby," Karch says. "Meanwhile, we'd made certain representations to the employees in the prospectus that we would close at a particular point, and the expiration date neared."

"Now Employee-Owned"

Parker and Karch were ultimately able to convince AWA to provide a bridge loan to close the deal on November 9, "much later than we had expected but still closure," Buth says. When the bond market regained health months later, the loan was paid off.

The day the deal closed, the company hung banners broadcasting the words "Now Employee-Owned" outside the plant. Employees celebrated in the parking lot by raising a flag with the letters "ESOP." It was an extremely emotional time. "We walked around thinking, 'We own the place now,'" says local union president Mickey Thompson. "This was made even better by the fact that we were busier than hell. There were high expectations for the future."

Charlie Boyd's company again was locally owned, for the first time in 30 years. On November 14, a full-page ad in *The Post-Crescent* announced: "We, the 2,600 employees of

Employees anticipated the day when they could affix the "Now Employee-Owned" banner above the entrance to the Appleton plant, above. Nearly 90 percent of company workers voted to contribute almost 75 percent of their retirement savings to buy stock in the company.

A new strategy also was in play. Buth planned to invest some of the company's cash flow to acquire companies outside its core enterprises but close enough to tap its intellectual capital, expertise and technology. "A lot of people were nervous about the Acquisition for Growth platform and wanted us to pay off our debt before spending money," he says. "But I said, 'We either reinvest or die.'"

Appleton Papers, have a stake in this great company. We are its owners. … We have roots that run deep into this community. We live here, we work here, and we believe in the bright future we have here." The pride in these words is evident.

Through the end of 2006, the return on employees' investment in their company has paid off handsomely, as the stock value has increased 236 percent.

Buth deserves a fair amount of credit for the successful buyout. "His strong leadership qualities were critical to the ownership change," Kathi Seifert, retired Kimberly-Clark executive vice president and an Appleton board member, told *The Post-Crescent*. "Doug was able to build trust with the employees and motivate them to pursue the buyout," she added. "It was a tremendous undertaking, and he was very successful in leading the company and the employees through [it]."

To better represent its diversified product set, Appleton Papers changed its name and logo in 2003, shortening the corporate name to Appleton and adding a tagline: "What Ideas Can Do."

A new strategy also was in play. Buth planned to invest some of the company's cash flow to acquire companies outside its core enterprises but close enough to tap its intellectual capital, expertise and technology. "A lot of people were nervous about the Acquisition for Growth platform and wanted us to pay off our debt before spending money," he says. "But I said, 'We either reinvest or die.'"

The board of directors acquiesced to Buth's plans, and in April 2003 Appleton acquired the first two of several new companies to join the fold — C&H Packaging and American Plastics, both based in northern Wisconsin. The deals marked Appleton's entry into the multi-billion-dollar performance packaging market.

American Plastics produces multilayered films and commercial packaging using a blown film coextrusion process in which melted resins are forced up a three-story tower through narrow slits in a ring-shaped die. The thin film takes the form of a tube or bubble and, as it cools, it is trimmed and rolled onto cores for use in commercial packaging.

C&H Packaging prints laminates and also converts flexible

C&H PACKAGING

C&H Packaging Company, Inc., of Merrill, Wisconsin, is one of two formerly privately held companies that Appleton acquired in April 2003 as part of a strategy to leverage its expertise in coating technology and microencapsulation in areas beyond paper.

"Producing substrates for the performance packaging industry is one of the markets in which we feel there are multiple opportunities for us to compete and to grow," then-CEO Doug Buth said in announcing the acquisitions. The other company acquired at the time was American Plastics Company, Inc., of Rhinelander, Wisconsin.

Founded in 1986, C&H employs about 100 people. The company develops, prints and converts flexible plastic packaging materials for companies in the food processing, household and industrial product and medical device industries. The company's innovative, stand-up pouches and plastic containers are used for food, drinks, snacks, dry mixes, pet treats and other perishable products.

C&H specializes in using coated materials to enhance the performance or add functionality to multilayered plastic films for these containers. The barrier properties of the coatings protect the contents from oxygen, odor, water and other contaminants. C&H develops packages from concept to completion and is especially adept at flexographic printing — a process that uses flexible plastic or rubber plates with raised surfaces and fast-drying inks to print on non-absorbent materials, such as plastic.

Pictured at left is C&H Packaging's laminations operation. Above, the company's packages are used to protect and preserve an array of perishable items.

AMERICAN PLASTICS

Appleton acquired American Plastics Company, Inc., of Rhinelander, Wisconsin, in 2003 as part of the company's Acquisition for Growth platform under CEO Doug Buth. The company also acquired C&H Packaging Company, Inc., of Merrill, Wisconsin, at the same time. The acquisitions enabled Appleton to enter the multi-billion-dollar packaging market, adding both technological capabilities and a new set of customers.

Founded in 1992, American Plastics has about 50 employees. The company produces high-performance barrier films and commercial packaging designed to extend the shelf life, ensure the integrity and increase shelf appeal of a variety of products. Meat, cheese, seafood, sauces, chemicals, automotive parts and tools are among the products sold in American Plastics' multilayered packaging — typically pouches.

American Plastics uses a process called blown film coextrusion to create plastic films with between five and eight layers. With on-site research and development, testing and prototyping, the company custom-develops packaging film with a variety of properties suited to unique performance requirements.

American Plastics converts some of its film products into pouches. Above, Dean Marquardt inspects a product on the company's pouch production line.

packaging for companies in the food-processing and industrial-products industries. The barrier properties of C&H's packages, which contain food, drinks, snacks and other perishable products, protect the contents from oxygen, odor, water and other contaminants.

Both companies, which collectively generated $40 million in net sales in 2002, are operated independently. "Some people saw the acquisitions as non-core, but I believed there were synergies in terms of opportunities to laminate film to paper," Buth says. "These were two great acquisitions complementing our areas of expertise."

The twin acquisitions were a warm-up. In December 2003 the company acquired BemroseBooth, a provider of secure specialized print services, for $60 million. Headquartered in Derby, England, BemroseBooth has a strong market position in security printed vouchers and payment cards, mass-transit and car parking tickets, variable data labeling, high-integrity mailings and printed calendars.

Another Acquisition

In 2004, on the 50th anniversary of the introduction of carbonless paper, Appleton took a moment to celebrate a technology that had changed the way business information was processed. Although the carbonless market was declining, there were still opportunities for new business. Such was the case when Hewlett-Packard, a major manufacturer of personal computers and printers, qualified Appleton's Xero/Form II carbonless sheet product for use on its new LaserJet printers. The decision marked the first time H-P had endorsed carbonless paper, reaffirming the continuing value of Appleton's signature product.

Doug Buth had thought of retiring at an early age, having lost friends in their 40s. Like many athletes, he was blessed with timing and knew it was best to leave at the top of his game. In April 2004, at the age of 49, he informed the Appleton board that he wanted to move on to other things.

Before ending his long career with the company, Buth piloted one more acquisition, the $68 million purchase of New England Extrusion, better known as NEX. Located in Turners Falls, Massachusetts, and Milton, Wisconsin, NEX produces single and multilayer polyethylene films for packaging applications.

American Plastics uses a blown film coextrusion process in which melted resins are forced up a three-story tower, left, through narrow slits in a ring-shaped die. The thin film takes the form of a tube or bubble and is cooled as it travels up the tower and is ultimately trimmed and rolled onto cores.

All in the Family

It's common in the paper industry for many employees to work alongside their siblings, parents and children. At Appleton this rich tradition has spawned a great number of multi-generational workers, men and women who can share more than a century's worth of tales about their company.

Take the case of Gary Koerner, an area electrician at Appleton today. Gary's grandfather, Elmer Koerner, joined The Appleton Coated Paper Company in 1912. "He was a supervisor in the finishing department and worked for the company until 1953," says Gary. "My dad, Robert, joined in 1955 as a trimmer operator and worked here until he died in 1994. My grandfather helped him get the job. One of my dad's favorite stories was when British American Tobacco bought the company. He and his buddies figured they'd now be able to get cigarettes at half price. No such luck, though."

Gary's daughter Whitnee worked at the company summers during her college years. "My wife Penny works here as a payroll specialist in HR," Gary says. "I even had an aunt work here, but I can't remember her name."

Four generations of Bert Wiegand's family also worked at the company. "My grandfather Martin Wiegand was a childhood friend of Mr. Boyd," Bert says. "In the 1920s he was briefly employed at the company until he was asked by Mr. Boyd to take care of his residences, which he did for many years. My father, Frank Wiegand, worked for the company before and after World War II, and his brother, my uncle Robert Wiegand, worked here for 35 years. I also put in 35 years, and my two sons Andy and Tim worked here several summers during their college years in the 1990s. Andy was a recipient of the company's Boyd Scholarship." Under the scholarship program, named for Appleton founder Charles S. Boyd, children or spouses of Appleton employees are awarded four-year grants of $2,000 per year to attend college.

Working for the company over many generations was not implausible, Bert Wiegand says. "We all grew up living less than a block away from company headquarters on Viola Street," he explains.

Matt Lyons' grandfather, Everett Lyons, worked at the Appleton plant in the 1940s "on one of the really old coaters," says Matt, an internal auditor who joined Appleton in 2006. "He got my dad, John Everett Lyons, a job here for a few months in the 1970s before he joined the military. My grandmother Lillian Schultz Lyons, Everett's wife, also worked here, and I have a cousin, John Wranosky, who works in maintenance. His son Jim works in the IS building."

Lydia Hoover, a buyer at the Roaring Spring mill, says her father-in-law, Ross Hoover, signed on in the 1940s at D. M. Bare Paper Company, which was purchased by the Combined Locks Paper Company in 1946. Her brother-in-law, Robert Hoover, worked in the machine room at the mill in the 1970s, and "his brother, my husband Gene, worked here for 38 years as a back tender on the No. 1 paper machine," Lydia says. "He started in 1964, and I started in 1971. Our two children don't work for the company, but Robert's son Mark works as a machine tender on the No. 3 machine."

The Bons family also has had four generations employed by the company, beginning with Matt Driessen, who worked at the Combined Locks mill as a millwright in the 1890s. "His son Henry Driessen got a job at 16 years old at the mill and retired in 1962 after 46 years' service," Linda Bons, an accountant for Appleton, says. "He was alive when I went to work for the company 27 years ago." Linda's dad is Robert Driessen, Henry's son. "Dad started work on November 11, 1950, and retired in 1989," Linda notes. "He was a foreman in the shipping and receiving department. I'm the first one to spend my entire career with Appleton, though two of my four children worked here during the summers. My mom and her father also worked at Locks. Her name is Rita Vandenburgt, and my dad's name is John."

There's a good chance that Linda's son Joseph may end up getting a full-time job at Appleton some day. "He's studying packaging engineering, and we're hoping he can find work here now that the company is in the packaging business," Linda says. "It runs in the blood."

Longtime Appleton employee Frank Sanders was in the third generation of his family to work for Appleton. His grandfather helped Charles Boyd install the company's first coaters and retired as Appleton's maintenance foreman, while his father (also named Frank) retired as personnel manager. Frank Jr.'s daughter, Chris Mueller, now works in human resources, and her son Andy and daughter Maggie had summer jobs at Appleton.

There are a number of multi-generational

families at the Roaring Spring mill, too. One of them is Rodney Claycomb, whose family connection dates back to mill worker Alfred James Pollard in 1843. As of 2006 more than 23 of Pollard's descendants have worked at the mill. Another is Greg Dick, who has a photo of his maternal grandfather, Emanuel Albright, "standing in front of the wood racks that used to be right off of the Sugar Street alley on Hogback in the late 1800s," he writes.

Heather D. Albright also lists five generations of her family working at the Spring Mill, beginning with her great-great grandfather, Levi Albright, and extending down through her great grandfather, Ross Decker Sr.; her grandfather, Charles Raymond Albright; her dad, Donald Ross Albright (who is still working for the company) and herself. "It is very possible that more generations have worked here," Heather writes. "The Albrights settled in this area when they first came from Germany in the 1800s."

She adds, "Thank you Appleton for the opportunity to carry on our family tradition."

Lloyd Swaim, left, Appleton's vice president of operations, bids farewell in 1970 to Cliff Hughes as Hughes retires after 43 years with the company. Cliff's father Owen, second from right, and brother Leo, right, also worked for Appleton for a total of some 60 years.

NEX complemented American Plastics in the performance packaging market. "Both companies produce film that is sold to converters that take this film and add value to it by printing on it and laminating it," says Kathy Bolhous, Appleton's vice president and general manager of performance packaging. "The film is then made into pouches used primarily for food products, as well as in the medical, health and beauty fields." Appleton's performance packaging companies are expected to enter new markets as the company leverages its experience in microencapsulation and coatings into new applications for polyethylene films.

Following the three acquisitions, Appleton had 3,400 employees, including 1,300 workers in the Fox Cities. While the diversifications broadened its scope, the company still manufactured a few legacy coated products, such as fluorescent, metallic, masking and durable papers. As in days of old, the company sold its products through merchants to printers. Appleton had assigned the trademarks for its Jazz and Currency papers in 2004 to CTI Paper USA, which enjoyed a strong sales and marketing reputation in the graphic arts community. The company also entered into a five-year agreement with Base Line, Inc. to distribute masking paper.

On his last day in office, Buth wrote an email message to employees reflecting on his time at Appleton. "The success of any business comes down to the people who own it and the people who work for it," he said. "Fortunately, at our company those groups are the same. … Thank you for investing in our company. We are the best possible owners."

Growing from the Core

Mark Richards took the baton in April 2005. The 45-year-old previously had been at Valmont Industries, where, as president of the engineered support structures division, he oversaw the design and manufacture of poles, towers and structures for lighting and traffic, wireless communication and utility markets. Under his care, the division's revenues grew an average 11 percent per year while achieving record earnings. Richards introduced a range of successful new products, increased the division's operating efficiencies and rode herd on several domestic and foreign acquisitions, making him the top candidate for a

BEMROSEBOOTH

In its 2004 acquisition of BemroseBooth, Appleton acquired a growing company in a set of markets that complemented Appleton's burgeoning security products business. The deal, which also bolstered Appleton's presence in the United Kingdom and Western Europe, was part of Appleton's strategy to offset the maturity of the carbonless paper market by participating in higher-growth markets that also leveraged company strengths.

At the time of the acquisition, BemroseBooth employed about 850 people at its manufacturing sites in Derby, Hull and Teesside, England and boasted sales revenue of approximately $100 million. William Bemrose founded the company in 1826 as a railway timetable manufacturer; the company sold its specialty print division and underwent a management buyout in 2000.

BemroseBooth's primary offerings include printed security and information products, mass transit and parking tickets, vouchers, payment cards, secure logistics and mail products, data labeling materials and promotional products. Customers include many of Western Europe's and the United Kingdom's biggest brands in retail, telecommunications, financial services and transportation.

BemroseBooth's products include security printed vouchers, above, and prepaid telephone cards produced on a Kora production line at its Derby facility, shown opposite.

Mark Richards, above, was the first Appleton CEO chosen from outside the company or one of its parents. He joined Appleton in April 2005 and quickly led the company to revisit its mission. The resulting strategy places renewed confidence in microencapsulation and coating technology as the fuel for growth in markets beyond those in which Appleton has traditionally participated. Multipart business forms, right and opposite, continue to serve as the primary application for the company's carbonless paper.

company like Appleton wanting to build new markets and extend older ones.

Richards was the first president and CEO of Appleton not chosen from inside the company or one of its parent companies. Soft-spoken, articulate and deeply thoughtful, he is known as a tenacious competitor. "He's the most disciplined person I know," his wife Jeanne told *The Post-Crescent*. She explained that at the age of 12, Richards was the worst player on his hockey team. He "wouldn't settle for that," practicing his stick handling and skating moves "for hours and days and weeks," she said. "He became a spectacular skater and next year was the most valuable player on his team."

In the months following his appointment, Richards led the company to adopt a new mission statement focusing the company on its technical papers and packaging businesses and placing a greater priority on serving international markets. To achieve these goals, he reorganized management, assembling around him a team of talented executives — and, once again, new blood has infused the company. The senior team's assignment is to execute a five-year strategic plan that calls for the company to become number-one or number-two in all of its markets, "targeting areas that offer the greatest growth and opportunity to win," Richards says.

Four Platforms

Richards has articulated four strategic platforms to assist this objective: Customer Focused Quality, Operational Excellence, Growth, and People Development and Diversity.

While continuing the emphasis on CFQ, Richards has been steering the initiative to make it more market-focused. "CFQ is all about using market insight to be better than anyone else at understanding our customers' needs," Richards says. To that end, Appleton is segmenting its customers to provide the highest possible levels of product quality, service, price and efficiency.

The Operational Excellence platform calls for putting in place measurable programs to continuously improve safety, quality, productivity and environmental management. It calls for the company to be world-class in protecting its workers as well as its communities and the environment, in offering customers the highest-quality products and services, and in being an efficient and cost-effective manufacturer. Appleton continues

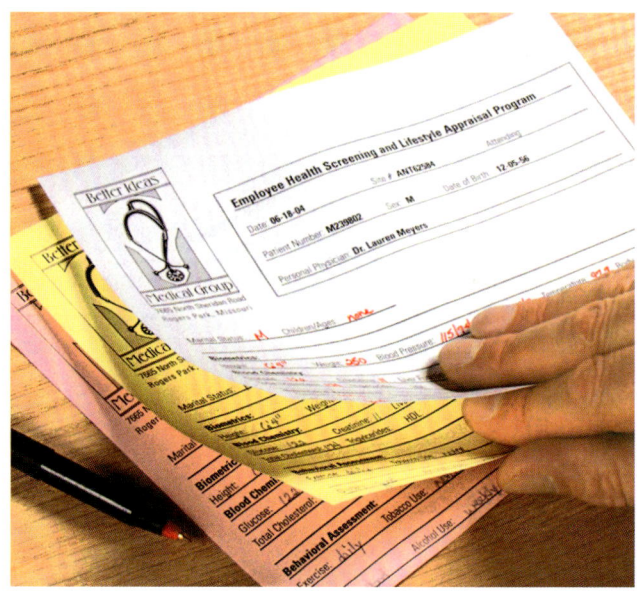

The company's thermal paper products are often used for transaction documents such as gas station and cash register receipts, coupons and, above, lottery tickets. Right, Senior Scientist Hugues Forget tests newly developed thermal products in a high temperature/high humidity chamber to ensure they meet resistance specifications required by Appleton customers.

Paper products requiring security features are a growing line of business for Appleton. The company can embed security features like "invisible" threads and strands, opposite, that can be seen only under a microscope, with fluorescent light or with scanning equipment. The critical nature of this work requires high-security status at the Roaring Spring mill, above, where a guard station is positioned to check the status of visitors.

to invest in lean manufacturing and Six Sigma initiatives to continuously improve operations, reduce waste and eliminate non-value-added costs.

As part of its Operational Excellence platform, the company named a chief compliance officer, Angela Tyczkowski, in 2006 to assess and manage the range of risks the company faces, including legal compliance, product, environmental, business continuity and operations. "Companies make money by taking risks and lose money by failing to manage them," she says, noting that Appleton's new focus on enterprise risk management intends to build a "risk-aware culture that takes and manages business risks wisely and deliberately."

The Growth strategy is based on identifying new business opportunities leveraging the company's existing products, markets and core competencies. Richards says Appleton will look to grow organically and potentially through acquisition — although, he says, any possible acquisition must "leverage one or more of our core competencies and extend our business into known markets or products." He also says the company will be "more aggressive in making decisions to exit businesses or drop projects that do not or will not produce the required return on investment."

The People Development and Diversity platform is designed to ensure ongoing development of future managers and leaders. It escalates the commitment to identifying, developing and promoting people with leadership potential in management as well as technical career tracks. It also establishes diversity and employee-development initiatives for people throughout the company, along with programs to improve organizational performance.

To assist Richards' growth-from-the-core strategy, Appleton restructured into three divisions: Technical Papers, Performance Packaging and International. The Technical Papers Division, combining the thermal paper and coated solutions businesses, focuses on three markets — high-volume transaction documents (carbonless rolls and point-of-sale thermal papers such as gas station and cash register receipts); low-volume/quick-response short-run papers (carbonless sheets) and security/entertainment products (airline and lottery tickets). "We have different markets and distribution channels so we've decided to align the businesses around them," says Kent Willetts, Appleton's vice president of marketing and strategy. "For example, low-volume/quick-response sells to paper merchants who sell to small local printers that do offset or digital printing, whereas carbonless rolls are sold to high-volume document processing companies."

Technical Papers

For the Technical Papers Division, Richards has established a growth strategy that calls for the company to leverage its core strengths to extend existing leadership positions as well as to enter new markets. "We have a franchise in carbonless and thermal where we lead the markets, and I firmly believe there are opportunities both internationally and in North America where we can build our market share," he asserts. "Our plan is to gain market share in carbonless. We also are looking to aggressively grow our international business, with a principal

Appleton's updated strategy calls for substantial growth to come from an expansion of the company's international business, driven especially in Latin America and Mexico. Appleton currently sells its technical papers products in 60 countries and created a new International Division in 2005 to pursue growth in Europe and Asia as well as in the Americas.

focus on thermal and a secondary focus on carbonless and security products. Already, we're growing double digits in our security line."

While the company's carbonless market may be mature, Richards and his team see Appleton's growth and future strength in the very technology and skills that built and sustained the carbonless business over a period of 50 years. He strongly believes carbonless offers significant revenue opportunities and is dismayed at the sometimes-voiced notion that this business is soon to be a backwater. "We are the market leader in carbonless and have a great opportunity to enlarge our position, picking up share from our competitors as they pull back or exit the market," he explains. "Moreover, there are many new markets we can enter leveraging our market-focused insight and the expertise we've accumulated in microencapsulation and innovative paper coatings through the decades."

In fact, Appleton's proprietary knowledge of the microencapsulation process and the scale at which the company can produce encapsulated materials serve as the foundation for launching a new business unit. The company has created commercially attractive ways to encapsulate fragrances and adhesives and is exploring ways to use microencapsulation, among a host of other ways, to enhance shelf-life and processability of certain substances, mask odors or tastes and protect unstable or sensitive materials. "This unique process that we have come to know so well over the past 50 years holds great potential to drive growth for our company through innovative applications," says Ted Goodwin, vice president of business development.

Performance Packaging

The Performance Packaging Division is focused on packaging applications in the food business that rely on its ability to provide high-quality custom films and commercial packaging. The division also looks to leverage the company's unique strengths — coating chemistry and microencapsulation — beyond paper and into plastic films. The division includes three of Appleton's subsidiaries: American Plastics, C&H Packaging and New England Extrusion.

"We are looking at specific niche areas of the market where we can become number-one or -two in market share," says

NEW ENGLAND EXTRUSION

Appleton acquired New England Extrusion, Inc., known in the marketplace as NEX, in 2005 — part of the company's long-term effort to accelerate the growth of its packaging business. With about 100 employees, NEX has two facilities, in Turners Falls, Massachusetts, and Milton, Wisconsin. It had sales of about $50 million at the time of the acquisition.

NEX manufactures polyethylene films of between one and three layers for high-performance packaging applications. NEX sells its film to converters, who add value to it by printing, laminating and/or making bags or pouches from it. Film produced by NEX typically ends up as packaging for food, personal care, medical and industrial products, including coffee, produce, frozen food, meat, cheese, snacks, pet food, direct-mail packages and lids.

NEX's business, markets and capabilities are similar to those of American Plastics, which produces high-barrier coextruded films of five to eight layers. In fact, Appleton specifically acquired NEX with an eye toward expanding the company's technical and manufacturing skills and product line and leveraging the two firms' complementary skills.

"Combining the capabilities of American Plastics and NEX will enable both companies to expand their abilities to design and produce film products with customer-specified properties," Appleton said in a news release at the time.

Appleton's performance packaging companies are expected to enter new markets as the company leverages its experience in microencapsulation and coatings into new applications for polyethylene films.

NEX manufactures monolayer and coextruded polyethylene film for demanding packaging applications.

Appleton's senior team: left to right, Angela Tyczkowski, vice president, secretary, general counsel and chief compliance officer; Kent Willetts, vice president of marketing and strategy; Sarah Macdonald, vice president and general manager of the International Division; Walter Schonfeld, president of the Technical Papers Division; Mark Richards, chairman, president and CEO; Kathy Bolhous, vice president and general manager of the Performance Packaging Division; and Tom Ferree, vice president of finance and CFO.

Kathy Bolhous, who leads the unit. "For example, we expect to become the market leader in higher barrier films — value-added film substrates that provide superior barrier protection for foods like beef jerky, which requires an 18-month shelf life. The five-year plan calls for growing revenue in the division by more than 10 percent annually.

International Division

Appleton sells its technical paper products into some 60 countries. In 2005 the company formed an International Division that includes a sales and marketing team focused on increasing those international sales. "We're on target to meet our performance expectations for 15 percent annual growth," says Sarah Macdonald, who heads the division as vice president and general manager. "We've positioned ourselves very well in geographic regions like Latin America and Mexico, where we have sales and distribution capabilities for label, point-of-sale thermal and carbonless products. Roughly 40 percent of our future international growth will come from this region, and we also have high expectations for growth in Asia and Europe."

Appleton's International Division also includes BemroseBooth, an Appleton subsidiary based in England that provides business-critical products with security features and specialized print management services to the transactional services, secure logistics and promotional products markets. Customers include many of the United Kingdom's biggest brands in retail, telecommunications, financial services and transportation. Appleton acquired BemroseBooth in December 2003.

Entering its second hundred years as a vibrant and resilient centenarian, Appleton is fueled by employee ownership and guided by energetic, smart and committed leaders — with employees and managers alike sharing a profound stake in the company's future. Today the small coating operation that opened for business in a converted pea canning factory is among the 20 largest paper companies in the nation — and one of the most profitable. In a testament to Appleton's stature, in November 2006 it for the first time made *Forbes* magazine's list of the country's largest private companies, placing 374th among the nearly 400 companies *Forbes* ranked by 2005 sales. Appleton had $1.05 billion in sales that year.

"Paper Bag Cholly" would be pleased to see the company he created, its values intact and employees still applying their ingenuity and effort to the demands of diverse markets. But Charlie Boyd would not be amazed. When he set out to build this company, he built it to last. Its resilience is a testament to the adroit skills of the company's 12 presidents and the thousands of other employees who have confronted extraordinary challenges with optimism and determination in quest of operational excellence and sustainable growth. As the company's mission statement posits, and Appleton's employee owners prove every day, "Passionate people believe in what they do."

Charles Boyd founded Appleton with the simple idea of coating paper to make it more printable. Entering its second century in 2007, Appleton was a vibrant, employee-owned company applying expertise in coating, encapsulation and security in a variety of traditional and new markets.

Timeline

1907
The Appleton Coated Paper Company is founded on May 13 in Appleton, Wisconsin.

1909
Company purchases the buildings it had been renting as its plant facilities for $4,000.

1911
Porcelain Enamel Shelf Paper is introduced.

1914
Main building for coating, calendering, cutting, sorting and trimming is completed at a cost of $44,000.

1920
New finishing building is erected for $54,000.

Company adopts group life insurance plan. Sales reach $2 million as company ships 6,800 tons.

1921
Appleton begins marketing coated papers under its own brand names to meet specific needs. Direct Sales Bond, company's first mill brand line, is introduced.

1922
Chief products are enameled book and magazine paper and lithograph labels.

1924
Company develops Coated Bond specialty product for use on gelatin roll duplicating equipment, first on the market.

1928
Woodbine Colored Enamel paper is introduced.

Largest building to date, containing the sorting, calender and sheeting operations, is completed.

1929
Institute of Paper Chemistry is founded.

1934
Company is founding member of Wisconsin Paper Group.

Appleton makes decision to concentrate production on specialty paper products.

1935
Air knife coaters begin to replace brush coaters.

Sales office opens in Chicago.

Currency Cover, a line of coated metallic papers for catalogs, booklets and annual reports, is introduced.

Supertuff Cover, designed for use in books with spiral and mechanical bindings, is introduced.

1936
Empress Offset Enamel, one of the first coated book stock papers for offset printing purposes, is introduced.

1944
U.S. Army purchases more than one million pounds of Supertuff for use in silk-screened outdoor signs and placards, and one million pounds of Woodbine Colored Enamel stock for masking purposes in the production of offset printing plates.

Cigarette "cup" paper business skyrockets, as company produces the paper for Lucky Strike, Pall Mall and Chesterfield brand packages.

1945
Fred Heinritz is named president; Charles Boyd remains chairman.

Major building projects include additions to the boiler house and main office building, construction of a laboratory and main office building, and building of five warehouses.

1946
Quarter Century Club is organized.

1948
Charles Boyd reassumes presidency upon the death of Fred Heinritz.

1950
New laboratory and mill office building, large finished goods warehouse, raw stock warehouse and receiving dock are completed.

Microencapsulation is invented by NCR chemists Barry Green and Lowell Schleicher.

1952

Richard Mahony is named president.

Charles Boyd dies.

1953

Coated two-side papers for cigarette packs represent almost half the company's total production, but decline precipitously in succeeding years.

The National Cash Register Company and Appleton conduct successful production trials of carbonless paper.

1954

NCR Paper brand of carbonless paper is introduced. Sales of carbonless paper "have greatly exceeded production," NCR states in its annual report.

1955

Suedetone greeting card paper is introduced.

For the first time, carbonless paper tonnage exceeds cigarette paper tonnage.

1957

Company celebrates semi-centennial, while city of Appleton celebrates its centennial.

Capital expansion in the last 10 years is five times that of the previous 40 years.

Non-carbonless paper products include battery paper and paper for bottle, can and phonograph labels.

1960

Former main office and company garage facing Meade Street are razed, and work begins on the largest plant addition to house new and future coating equipment.

1961

New high-speed, aqueous coater is installed in new building at a cost of $5 million.

1962

John Reeve is named president.

1963

Property is purchased to provide space for experimental solvent coater.

Thermal paper manufacturing is introduced.

1964

Company receives permission to move Meade Street 150 feet to accommodate plant expansion.

1965

Installation and startup of a second high-speed, electronically controlled coating machine are accomplished.

1966

Four acres for future expansion are purchased.

1967

Company installs world's first tandem coater to manufacture carbonless paper.

Appleton establishes major research and marketing center.

William Siekman, son-in-law of Charles Boyd, is named company chairman.

1969

NCR acquires The Combined Paper Mills, which owns the Combined Locks paper mill in Wisconsin and Roaring Spring mill in Pennsylvania.

1970

Ascot, the company's first non-paper, coated plastic sheet product, makes its debut. Product uses Dupont's Tyvek.

NCR buys Appleton and, in 1971, unites it with NCR subsidiary Combined Mills to form Appleton Papers, Inc., with headquarters in Appleton.

1971

Carbonless sales, marketing and technical groups begin gradual migration from NCR in Dayton to Appleton.

Company begins converting paper machines at Roaring Spring mill to make carbonless paper.

Non-carbonless microencapsulation products like "scratch and sniff" paper, encapsulated perfume on cash register paper, time-release aspirins and mood rings are introduced.

1973

Thermal chart and calculator papers are introduced.

New plant in Harrisburg, Pennsylvania, begins operations.

1974

New colors, including canary and pink, are added to carbonless product line. Also, for the first time, company begins making carbonless capsule dyes that produce a black image.

1976

New high-speed coater is installed at Appleton plant with the potential to coat carbonless paper at twice the previous speed and output.

1977

John Hangen is named president and CEO.

Total sales of NCR brand carbonless paper since its introduction exceed one million tons.

1978

NCR sells Appleton to B.A.T. Industries.

1980

Company forms subsidiary in Canada to market products made in the United States.

1982

Tom Busch is named president; John Hangen remains CEO.

1983

Thermal fax paper for document transmission over phone lines is introduced.

1984

Company purchases a paper converting plant in Peterborough, Ontario, to meet regulations for doing business in Canada. The facility is sold in 1989 when regulations change.

Company purchases P. H. Glatfelter paper and pulp mill at West Carrollton, Ohio, for $83 million.

Tom Busch is named CEO.

1985

Appleton launches quality improvement process.

1986

John Turner is named CEO. Dale Schumaker becomes president.

Company purchases East Shore Chemical Company in Muskegon, Michigan, a manufacturer of black dye for microcapsules.

New corporate theme is unveiled: "Think Quality — do it right the first time."

1987

West Carrollton mill restarts No. 93 paper machine for carbonless paper production.

1988

No. 13 thermal coater comes online at Appleton plant.

1989

Company expands its nationwide distribution network to provide one-shipment delivery of any order confirmation.

Pearlescent grade is added to the Currency Cover line.

Gordon Bond is named chairman and CEO.

1990

Company introduces first recycled carbonless paper called Recover and unveils Foampak recycling concept.

De-merger of Appleton is created by the breakup of B.A.T. Industries. Appleton is joined with U.K.-based Wiggins Teape to form Wiggins Teape Appleton.

Wiggins Teape Appleton merges with French paper group Arjomari-Prioux to become Arjo Wiggins Appleton, or AWA.

1991

Appleton sells East Shore Chemical Company to Mitsui Toatsu Chemicals and its affiliate Yamamoto Chemicals.

Appleton begins $170 million expansion of the Combined Locks mill.

1992

Appleton becomes the first manufacturer to achieve ISO 9002 certification for the production of thermal paper.

Company unveils Xero/Form sheets for high-speed copiers and laser printers.

1993

Dale Schumaker is named CEO.

Company begins $35 million thermal paper expansion of Appleton Plant.

Company convenes first Conference for Quality Achievement at Roaring Spring.

1994

Company mounts unsuccessful attempt to acquire S. D. Warren, a coated free sheet producer.

Docucheck "fraud-resistant" security papers are introduced.

Appleton receives a site visit by examiners for Malcolm Baldrige National Quality Award, and is a finalist in the competition.

Company completes installation of No. 7 paper machine at Combined Locks.

1995

Appleton enters the coated free sheet market with the purchase of a paper mill in Newton Falls, New York, and further expansion of the Combined Locks mill to accommodate coated free sheet paper production.

Richard Curwen is named CEO.

Tom Busch is inducted into the Paper Industry International Hall of Fame.

1996

Appleton is again a finalist in Baldrige Award competition.

1997

NCR scientist Barry Green, co-inventor of microencapsulation, dies.

1998

Appleton's Year 2000 (Y2K) compliance project becomes part of the Smithsonian Institution's permanent research collection on information technology innovation at the National Museum of American History.

Doug Buth is appointed CEO and president.

1999

West Carrollton mill earns ISO 14001 certification.

Appleton closes Harrisburg plant and moves two coaters to the Appleton plant. Company sells the Harrisburg plant in 2001.

2000

Appleton exits coated free sheet market when its parent company, AWA, transfers the assets and operations of its coated free sheet business to a new legal entity, Appleton Coated LLC in Kimberly, Wisconsin.

2001

Roaring Spring mill earns ISO 14001 certification.

Charles S. Boyd is inducted into the Paper Industry International Hall of Fame.

Appleton employees complete $810 million buyout of the company.

2003

Appleton Papers changes its trade name and logo to simply Appleton, with a new tag line: "What Ideas Can Do."

Appleton enters the flexible packaging market with the acquisition of C&H Packaging and American Plastics.

Appleton acquires BemroseBooth, a leading British provider of secure and specialized print services.

2004

John Reeve is inducted into the Paper Industry International Hall of Fame.

2005

Appleton acquires New England Extrusion, a producer of single and multilayer polyethylene films for packaging applications.

Mark Richards is named CEO and president.

Company is organized into three divisions: Technical Papers, Performance Packaging and International.

Appleton wins Company of the Year award from *Solutions* magazine.

2006

Appleton is included on the *Forbes* list of the country's largest private companies.

2007

Appleton celebrates its 100th anniversary.

Appleton's Chief Executive Officers

Charles Boyd
1907-1945, 1948-1952

Fred Heinritz
1945-1948

Richard Mahony Sr.
1952-1962

John Reeve
1962-1977

John Hangen
1977-1984

Tom Busch
1984-1985

John Turner
1986-1989

Gordon Bond
1989-1993

Dale Schumaker
1993-1995

Richard Curwen
1995-1998

Doug Buth
1998-2005

Mark Richards
2005-present

Acknowledgments

About the Author:

Russ Banham is a veteran business journalist and the author of 12 books, including the best-selling *Rocky Mountain Legend*, the story of the Coors brewing dynasty; *The Ford Century*, the award-winning centennial history of Ford Motor Company; *Wanderlust: Airstream at 75*; and his latest, *Twenty Years of Independence: The Anadarko Story*. His television appearances include NBC's *Today* show and *Biography* on the A&E channel.

Acknowledgments:

I would like to thank the people of Appleton who were so kind to me during my stays in their delightful city, as well as The History Museum in Appleton and the Paper Industry Hall of Fame, which gave me private tours, flattening my learning curve about the paper industry and the city's history. I would also like to acknowledge all the current and former employees of Appleton who shared their memories with me. In particular, I want to thank Dan McIntosh, who was always there to answer a question or check a fact for accuracy. Kudos also to Bill Van Den Brandt, who helped me understand the company, opened my eyes to a treasure trove of historical data and was my courteous and knowledgeable escort during my visits. Big thanks are also due my research assistant, Jennifer Sue Johnson, who provided countless hours of work toward assembling the best history possible.

Index

Bold listings indicate illustrations.

A

Adding Value to Paper, 56
Adstaff, 91
advertising, 34, **34**, 35, **35**, 36, **36**, 46, **46**, 60, 61, **61**, **62**, 63, **63**, 86, 88, **88**, 89, **89**
Albright, Charles Raymond, 121
Albright, Donald Ross, 121
Albright, Emanuel, 121
Albright, Heather D., 121
Albright, Levi, 121
Alco Standard, 100
Allyn, Arthur, 80, **81**
American Can Company, 72
American Color Trends, 64
American Envelope Company, 99
American Plastics Company, Inc., 117, 118, **118**, 119, **119**, 123, 131
American Tobacco Company, 45, 50, 120
Apco Hi-Fold Cover, 64
Apco Masking Paper, 64
Appleton (city), 10, 11, **11**, 12–13, 19, 67
Appleton Coated LLC, 106
Appleton Coated Paper Company
    anniversaries of, 64, 67
    calendering department of, **18–19**, 19
    cutter room, **20**, 21
    establishment of, 15, 16
    expansion, 21, 26, 29, 42, 46, 64, 67, 71, 76
    finishing room, 27, **27**
    incorporation of, 17, **17**, 19
    international sales, 28
    landscape/garden, **14**, 15, 38, 67, **67**
    logos, 90, **90**
    NCR purchase of, 76, 82
    office building, 16, **16**
    plant, 21, **24**, 25, **25**, 48, **48**, **70**, 71
    production, 42, 48
    research and development, 25, 46, **46**, 76
    restructuring, 72, 76
    on selling, 77, 80
    strategic meetings, 71
    See also Appleton Papers Inc.
Appleton Crescent, The, 10
Appleton Enamel, 37
Appleton Label Plate, 36
Appleton Papers Inc.
    acquisitions, 100, 106, 117, 118, 123, 128, 132
    AWA wanting to sell, 106, 110
    employee buyout of, 110–115, **110**, **113**, **115**, 117–118
    expansion, 82, 86, 95, 99, 100, 101, **101**, 102–103
    formation of, 69, 80
    international sales, 124, 128, 130
    logos, 90, **90**
    name changes and, 91, 117
    operated by BATUS, 95
    performance packaging market and, 117
    plant, 82, **82**, **83**, 86, **86**, **87**, 89, **89**, 92, **92**, **93**
    quality program, 96, 100, 102, 103, **104**, 105, 106, 110
    restructured into three divisions, 128–132
    stock value, 117
    strategic business units, 102
    strategic platforms of, 124, 128
    See also Appleton Coated Paper Company
Appleton's Vocational School, 33
Apro, 45
Arent, Kerry, 110–111, **111**, 113
Arjomari-Prioux, 102
Arjo Wiggins Appleton (AWA), 102, 106, 110–111, 112, 114, 115
Ascot, 84, **84**, 85–86
Ashman, Henry, 45, **45**

B

Babcock, Harold, 65, **65**
Badger Periodicals, 46
Bank Equipment News, 63
Bankers Monthly, 63
Bare, D. M., 74
Base Line, Inc., 123
B.A.T. Industries, 90, **90**, 91, 95, 102
BATUS, 95–96, 106
Baum, Henry, 95
Bear Stearns, 111, 114, 115
BemroseBooth, 118, **122**, 123, **123**, 132
Benefit Consultants Incorporated (BCI), 112, 113, 114
Bennett, Joan, 92
Berge, Herman B., 42, **43**, 45, **45**
Berghuis, Barnard, 69
Berghuis, George John, 69
Berghuis, Johanna Welhuis, 69
Bergstrom Paper Company, 19, 35, 42, 64, 96, 99
Bernstein, Mark, 111, 114
BIFI (Bright Ideas for Improvement), 103
Birren, Faber, 64
Blue Mellochrome Dull Coated Bristol, 64
Bodmer, Karl, 11
Boise Cascade, 77, 100, 102
Bolhous, Kathy, 132, **132**
Bond, Gordon, 102, **102**, 103, 105
Bons, Joseph, 120
Bons, Linda, 120
Bowlby Business College, 29
Boyd, Bertha, 17
Boyd, Charles Samuel, 12, 13, **13**, 15, **15**, 17, 19, 21, 22, 24, 25, 26, 28, 29, 32, 33, 35, 37, **37**, 38, 42, **44**, 45, **45**, 48, 50, 56, 67, 120, 133
Boyd, Cornelia S. (Bowen), 15, 17, 19
Boyd, Florence, 17
Boyd, Major Thomas, 15
Boyd, Martha, 15, 22, 26, 48
    See also Siekman, Martha Boyd
Boyd, Robert, 17, 19, 26, 28
Boyd, Samuel, 12, 17
Boyd Scholarship, 120
Braun, Rick, 113
British American Tobacco Company, 91
Brown and Williamson Tobacco Company, 95
Brown, John, **20**, 21
Buchanan, William E. "Bill," 80, **81**
Busch, Thomas, 12, 46, 56, **56**, 60, 62, 64, 69, **69**, 71, 76, 77, 80, **81**, 82, 88, 96, 99
Buth, Douglas, 106, 110, 111, **111**, 113, 114, **114**, 115, 116, 117, 118
Bynum, Curtis, 17, 19

C

C&H Packaging Company, Inc., **116**, 117–118, **117**, 131
Calmes Carriage Works, 21
Calmes, Frank, 21
Canada, 95–96
carbonless paper, 62, 65, 67, 71, 74, 76, 77, 80, 82, 84, 85, 91, 92, 95, 96, 99, 100, 103, 105, 110, 118, **123**, **124**, **125**, 128, 131
    advertising, 58, 59, **59**, 60, 61, **61**, 62, 63, **63**, 86, 88, **88**, 89, **89**
    beginnings/development of, 56, **56**, 58–59, **59**, 60–62, 64
    See also microencapsulation technology
Carbonless Roll Attribute Study, 96
casein, 21, 22, 26, 38, **38**, 42, 45, 58, 60
Casey, Larry, 72, 76, **76**, 85
Cashman, Tom, 86, 91, 100, 106
Chady, Lee, 45, **45**
Champion, 77
Charles S. Boyd Paper Company, 15, 17, 19, 22
cigarette packaging, 37, 42, **43**, 45, 48, 50, 52, 74, 77
Clampitt, 88
Clark, Andy, 65, **65**
clay, 22, 26, 50, 58, 60
Claycomb, Rodney, 121
coated free sheet business, 90, 106, **107**, 110
coated paper, 15, 17, 19, 22, 26, 28–29, 36–37, 67, 84, **84**, 92, **92**, 102, 106, 110, 123, 128, 133
    See also names of individual coated paper products
Colbert, Claudette, 92
Collar, Milt, 66, **66**
color department, 38, **38–39**
Colvin, Gene, 37, 50
Combined Locks mill, 85, 86, 91, 95, 102, 103, 106, 120
Combined Locks Paper Company, 61, 62, 69, 71, 74, 76, 120
Combined Locks, village of, 69
Combined Paper Mills, 76, 80, 85
Computer Data Center, 100
Conference for Quality Achievement, 103, **104**, 105
Consolidated Paper Company, 50, 77
Copco, 88
Coyle, Don, 66, **66**
CTI Paper USA, 123
Currency Cover, 37, 76, 92, 123
Curtis, Rick, 110
Curwen, Richard, 106
Customer Focused Quality (CFQ), 103, 105, 106, 110, 124

D

Damadaran, Satish, 99
DeBruin, Quentin, 67, **67**
Decker, Ross, Sr., 121
Deep Blue System, 86
DeGroot, Sy, **20**, 21
Deloitte LLP, 111
Denstedt, Ethel, 45, **45**
Dick, Greg, 121
Diem & Wing, 88
Direct Sales Bond, 28, **28**, 29
D. M. Bare Paper Company, 74, 76, 120
Dorschner, Vilas, **44**
Driessen, Henry, 120
Driessen, Matt, 120
Driessen, Robert, 120
Driscoll, Kelly, 112–113
Dullchrome Coated Book, 36
Duplex Coated Paper, 35–36
DuPont, 51, **84**, 85
Dwight Brothers Paper Company, 22

E

Eastman Kodak, 51
East Shore Chemical Company, 100, 102
Eby, John, 74
Eggert, Edna, 67, **67**
Eggert, Max, **44**
Elkins, Stephanie Essex, 114
Employee Retirement Income Security Act, 112
employees, 16, **16**, 19, 22, **23**, 26, **32–33**, 33, 41, **41**, 45–46, **47**, 48, 64, **65**, 103
    multi-generational, 120–121
    number of, 21, 22, 27, 29, 42, 45, 48, 64, 76, 91, 103, 110, 123
    ownership of Appleton Papers Inc. by, **110**, 110–115, **113**, **115**, 117–118
    recreational events for, 67
    sports teams, 26, 27, **27**, 64, 66, **66**, 67, **67**
    See also Quarter Century Club; unions
Empress Offset Enamel, 37, 64
environment, 86, 96, **97**, 99, 112, 124, 128
Ernst, Marvin, 49, **49**
ESOP (employee stock ownership plan), 110, 111, 112, 113, 115
Eve, 86

F

Farm Guest House, 69, **69**
Farquhar, William, Sr., 45, **45**
Feistel, Sally, 113, 114
Ferber, Edna, 10, 13
Ferguson, Mark, 96
Ferree, Tom, 132, **132**
Ferrell, Ed, 66, **66**
Ferry Morse Seed Company, 50
festooning process, 22, 46
50 Colorful Years, 29, 37, 64, **64**
First National Bank of Appleton, 19
Foampak recycling program, 96, **96**, **97**
Forbes magazine, 133
Forget, Hugues, 126, **127**
Fox Heritage, The (Kort), 10
Fox River, 10, **10**, 11, **11**, 21, 53, 69, 112
Fox River Paper Company, 80
Fox Tractor, 71
Fox Valley Technical College, 33
Frinak, Sherm, 50, 71, 82

G

Galpin, Alfred, Jr., 13
Gauerke, Anton, 45, **45**
George A. Whiting Paper Company, 12, 17
Georgia Kaolin, 50, 51
Gimbels, 95

Glatz, Roland, 66, **66**
Glidden Company, 42
Gochnauer Concrete Products, 46
Godfrey & Kahn, 112
Goetz, Bill, 72, **73**, 88
GO! (Growth Opportunities), 110
Goldsmith, Sir James, 102
Goodwin, Ted, 131
Gosz, Bernice, 49, **49**, 67, **67**
Great Depression, 29, 32–33, 35, 42
Green, Barrett K. (Barry), 50, 51, **51**, 52, 53, 58, 59, **59**, 60
Growth platform, 124, 128

H

Haddad, Leonard, 80, **81**
Hall Paper, 88
Hangen, John, 91–92, 95, 96
Hayton Pump and Blower Company, 19, 29
Heinritz, "Bud," 45, 46, 48
Heinritz, Fred, 22, 26, 29, **29**, 33, 42, **44**, 45, **45**
Helser, Earl, 45, **45**
Helser, Lester, 45, **45**
Herbig, Jim, 85
Hewlett-Packard, 118
Hiler, Jerry, 66, **66**
Hilton Davis, 100
Hofferberth, Herb, 88
Holtz Nursery, 46
Hoover, Gene, 120
Hoover, Lydia, 120
Hoover, Mark, 120
Hoover, Robert, 120
Hoover, Ross, 120
Houdini, Harry, 12, **12**, 13
Houlihan, Lokey, Howard & Zukin, 110
Hudson, Ray, 91
Hughes, Cliff, 121, **121**
Hughes, Leo, **38**, 121, **121**
Hughes, Owen, 121, **121**
Hurley, Cliff, 65, **65**

I

information technology, 100, 102, 110
Ink Color Guide (Birren), 64
Institute of Paper Chemistry, 12, 29, 33, 35, 36, **36**, 50
International Business Machines (IBM), 15
International Division, 132
International Paper Company, 51, 52, 91, 100
international sales, 28, 62, 124, 128, 130
ISO 14001 registration, 99

J

Jazz, 123
Jensen, Mike, 92, **93**
Jezerc, Ron, 72, 76, 82, 85, 86, 88
John Leslie Paper Company, 22
Jolliet, Louis, 53
Just-In-Time principles, 103

K

Kahler, Emil, **44**
Kampo Transport, 64, 65, **65**
Kansas City Paper House, 22
Karch, Paul, 111, **111**, 112, 113, 114, 115
Kimberly-Clark Corporation, 12, 19, 24, 50, 77, 96, 99
Kircher, Helton and Colett, 60
Kneepkens, Al, 46, 66, **66**
Koerner, Elmer, 45, **45**, 120
Koerner, Gary, 120
Koerner, Penny, 120
Koerner, Robert, 120
Koerner, Whitnee, 120
Koletzke, Cora, 45, **45**
Kort, Ellen, 10
Kubitz, Fred, 45, **45**
Kula, Steve, 111, **111**, 112

L

Lang, Ted, 45, **45**
Lauer, Howard, 52, 53, 56, 76, 80
Lawrence, Amos, 10, 12
Lawrence College/University, 12, **12**, 13, 15, 17, 22, **22**, 29, 32, 46, 67
Lewenstein, Abe, 42, 46, 56, **56**, 76
Lin, Charlie, 88
Litho Label, 36, 37, 48
Lowe, John, 17, 19
Luebke, Orv, 48
Lust, Carl, **39**
Lyons, Everett, 120
Lyons, John Everett, 120
Lyons, Lillian Schultz, 120
Lyons, Matt, 120

M

Macdonald, Sarah, 132, **132**
Mac Group, The, 102
Mahony, Richard, 26, 29, 33, 38, **38**, 48, 50, 51, 52, 56, 71, 76, 80, **81**
Malcolm Baldrige National Quality Award, 102, 103, **103**, 105, 106
Marathon Corporation, 72
Marquardt, Dean, 118, **118**
Marquette, Father Jacques, 53
Masterfold Enamel, 37
McCann, Paul, 21, 45
McIntosh, Dan, 85, 92, 96, **96**, 100, 103, 105
McKinsey & Company, 72
Mead Paper Company, 50, 61, 62, 71, 76, 82, 85, 91, 96, 100
Medical Economics, 63
Meier, Paul, 91, 95, 103, 105
Mellochrome Dull Coated Bristol, 36, 64
Mendels, Ed, 86
Meyer, Roland, 56, **56**
Mickelson, George, 56, **56**
microencapsulation technology, 50, 51, 52, **52**, 53, **53**, 56, 58, 72, 85, 86, 88, 95, 124, 131, 133
   See also carbonless paper
Mielke, Eva, **44**
Milwaukee Journal Sentinel, 113, **113**
Minton, Ken, 110
Mittelstadt, Lloyd, 38–39
Modern Hospital, 63
Moll, Gus, **20**, 21, 45, **45**
Moll, John, 45, **45**

Monsanto, 51
Montana, Joe, 110
Montello, Ralph, 53
Moore Corporation, 71, 80
Morrison, Helen, 111
Morrison, John, 74
Moser Paper Company, 17, 28
Mouth of the River (Bodmer), 11, **11**
MRP (Manufacturing Resource Planning), 100
Mueller, Andy, 120
Mueller, Chris, 120
Mueller, Maggie, 120
Muenster, Edward, 45, **45**
multi-part business forms, 124, **124**, **125**
Munder, Norman T. A., 34
Mundt, Ray, 100
Myers, Vance, 74

N

Nashua Corporation, 95
National Cash Register Company (NCR), The, 50, 51, 52–53, 56, 60, 69, 71, 76, 77, 80, 82, 85, 88, 91
   See also carbonless paper
National Office Paper Recycling Project, 96
National Recovery Administration (NRA), 36–37
NCR Paper brand. *See* carbonless paper
Neenah Paper Company, 12
Nekoosa-Edwards Paper Company, 62, 64, 71, 76, 77
New England Extrusion, Inc. (NEX), 118, 123, 131, **131**
Newton Falls mill, 106
Nicolet, Jean, 10
Northern Transportation Company, 46
Nowak, Frank, 45, **45**
Noyes, Chris, 111–112, **111**, 114, 115

O

Occupational Safety and Health Administration (OSHA), 86
Oelman, Robert S., 69, 80
Office and Reproduction Review, 63
Olin, 77

Operational Excellence platform, 124, 128
Oshkosh, Chief, 10
Otto, John, 45, **45**
Ownership Updates, 112

P

Paden, Keith, 74, 99
Page, Bill, 56, **56**, 76
Paone, Lou, 110, 111, 113
Paper Industry International Hall of Fame, 12, 77
Parker, Dale, 111, **111**, 115
Paveri-Fontana, Luca, 110
Peculiar Treasure, A (Ferber), 13
People Development and Diversity platform, 124, 128
Peotter, Edward, 45, **45**
Performance Packaging Division, 131–132
Permakolor Litho Label, 37
Peterson, Pike, 100
P. H. Glatfelter Company, 77, 96, 99
Philip Crosby Associates, Inc., 96, 105
Philip Morris, 77, 80
Phillips, Jim, 115
Phillips, Paul, 82, 86, 88
Pollard, Alfred James, 121
Polychrome Dull Coated Book, 29, 34, **34**, 36, 37
Post Card, 36
Post-Crescent, The (Appleton), 22, 26, 37, 42, 64, 115, 117, 124
Potlatch, 77
Principal Financial Group, 112
Prodoehl, Cindy, 113, 114
Prodoehl, Pete, 112, 113, 114
Project Venture, 110

Q

quality. *See* Customer Focused Quality (CFQ)
Quarter Century Club, 44, **44**, 45, 67

R

Radke, Harold, **39**
Rally brand, of carbonless paper, 95
Ray-O-Vac, 50

RCA, 72
recycling, 96, **96**, 97
Reeve, Dr. John T., 13, 24
Reeve, John, 12, 15, 21, 24, 26, 32–33, 37, 38, 42, 45, 48, 50, 61, 64, 71, **71**, 72, 76, 77, **77**, 80, **81**, 91
Reflect-O-Ray, 50
Reformed Temple Zion, 12
Regenfuss, George, **44**, 65, **65**
Reinhard, Don, 95
rewinder department, 22
Richards, Jeanne, 124
Richards, Mark, 99, 123, 124, **124**, 131, 132, **132**
Richmond Brothers Paper Mill, 10
R. J. Reynolds, 85
Roaring Spring mill, 74, 75, **75**, 80, 85, 86, 91, 95, 99, 102, 103, 105, 106, 113, 120, 121, 128, **128**
Roche, Pat, 66, **66**
Rogers, Earl, 45, **45**
Rogers, Henry James, 13
Rollo, Wanda, 67, **67**
Roosevelt, Franklin D., 36, 37
Rouman, Art, 76
Ruby (Charlie's secretary), 42
Russler, D. W. "Russ," 69, 72, 77, 80, **81**

S

safety, 38, 86, 124
Saks Fifth Avenue, 95
Sandberg, Bob, 52, 56, 60
Sanders, Frank A., 15, 33, 45, **45**, 52, 64, 92, 120
Sanders, Frank E., 33
Satorius, Stan, 66, **66**
Schabow, Armin, 66, **66**
Schinke, William, **20**, 21
Schleicher, Lowell, 51, **51**, 52
Schmidt, Bob, 65, **65**
Schonfeld, Walter, 132, **132**
Schulz, Don, 66, **66**
Schumaker, Dale, 21, 69, **69**, 71, 72, 76, 88, 91, 95, 96, 100, 102, 103, **103**, 106
Schwab, Helmut, 58
Science Research Associates, 15

S. D. Warren, 50, 76, 106
security products, 74, 128, 131, 133
Seidl, Joyce, 67
Seifert, Kathi, 117
Sensenbrenner, Frank J., 24, 26
Sensenbrenner, Howard, 26
Siekman, Bill, 15, 22, 35, 38, 42, 76, 77, 80, **80**, **81**, 85
Siekman, Martha Boyd, 80, **81**
    See also Boyd, Martha
Silvercote, Inc., 50
Six Sigma initiatives, 128
Smith, Karlton, 99
Smith, Lloyd, 45, **45**
sorting department, 65, **65**, 76
space utilization project, 102
Spangs Mills, 74
specialty papers, 29, 34, **34**, 35–36, 37, 64
    See also coated paper
Spencer, Lois Hill, 15, 26
Spencer, Lyle, 15, 26, 32
Stage, Al, 66, **66**
Standard Manufacturing Company, 76
Stark, Carlton, **39**
State Street Global Advisors, 112
Stevens, James A., 49, **49**
Stiltjes, Ivan, 65, **65**
Stora AB, 106
Suedetone, 64
Suess, Bob, 72, 91
Sullivan, Tom, 71, 88, 91, 99
supercalender, 17, 19, 21, 22, 46
Supertuff Cover, 37, 45, 64, 76
Swaim, Lloyd, 72, 76, 80, 121, **121**
Sylvester, George, 45, **45**
Systemedia, 91

T

Teacher's Pet, 85
Technical Papers Division, 128, 131, 132
thermal paper, 71–72, 76, 80, 88, 92, **94**, 95, 96, 99, 101, 102, 103, 126, **126**, 128, 131
Thiel, Jerry, 86, 92
Thies, Louis, 45, **45**
Thompson, Mickey, 76, 88, 115
3M Company, 71
Tobacco Securities Trust, 91
Topeka Paper Company, 22
trimming department, 22, **40–41**, 41
Truttschel, Paul E., 50–51, **50**, 52, 56, 76, 80
Turner, John W., 96, 100, 102, 105
Tyczkowski, Angela, 128, 132, **132**
Tyvek, **84**, 85, 86

U

unions, 42, 71, 76, 88
Union Carbide, 50
Unisource, 100
United Mine Workers, 71
United Paperworkers International Union, 76
United Steel Workers, 76
University of Wisconsin, 48
U.S. Department of Commerce, 103, 105
U.S. Patent Office, 51
U.S. Securities and Exchange Commission (SEC), 111, 112

V

Valmont Industries, 123
Vandenburgt, Rita, 120
Van Eyck, Tim Jones, 92, **93**
Van Wyk Coal Company, 46
Villa Woodbine, 15

W

Watson, Thomas, Sr., 15
Wehrel, Dick, 110, 111, **111**, 114, 115
Weiss, Erich, 12
    See also Houdini, Harry
Weiss, Rabbi Mayer Samuel, 12
Wellness Center, 21, 77, **77**
West Carrollton mill, 96, **98**, 99, **99**, 102, 103
Westphal, Chester, 65, **65**
Weyenberg, Matt D., 26, 33, 45, **45**
Weyerhaeuser, 77

Whalen, Ann, 110
Wheeler, Richard, 45, **45**
Wheelock, Woodward, **39**
Whispering Pines School, 46
White Porcelain Enamel Shelf and Lining Paper, 22
Wiegand, Andy, 120
Wiegand, Bert, 120
Wiegand, Frank, 120
Wiegand, Martin, 120
Wiegand, Robert, 120
Wiegand, Tim, 120
Wiggins Teape Appleton, 102
Wiggins Teape, 62, 91, 95, 102
Wikoff, Jack, 61–62
Willamette Management & Associates, 113
Willetts, Kent, 128, 132, **132**
Winnebago Indians, 10, **10**
Wisconsin Department of Financial Institutions, 114
Wisconsin Enamel, 37
Wisconsin Methodist Church, 12
Wisconsin Paper Group, 35
Woodbine Colored Enamel, 28–29, 37, 45, 46, **46**
Woodbine Duplex Enamel, 36, 64, 76
Woodbine Folding Enamel, 37
World Trade Center terrorist attacks, 114–115
World War II, 42, 45, 46, 49
Worms & Cie, 106
Wranosky, Jim, 120
Wranosky, John, 120

X

Xero/Form II carbonless sheet, 118
Xerox, 72, 105

Y

Yamamoto Chemicals, 100, 102

Z

Zeffery, Florian, 45, **45**
Zellerbach Paper Company, 88, 96, 100
Zurilla, Althea, 67, **67**